TETRAHEDRON ORGANIC CHEMISTRY SERIES

Series Editors:

J-E Bäckvall
University of Stockholm,
10691 Stockholm, Sweden

J E Baldwin, FRS
University of Oxford,
Oxford, OX1 3QY, UK

R M Williams
Colorado State University,
Fort Collins, CO 80523, USA

VOLUME 24

Molecular Diversity and Combinatorial Chemistry

Principles and Applications

Related Titles of Interest

BOOKS

Tetrahedron Organic Chemistry Series:
CARRUTHERS: Cycloaddition Reactions in Organic Synthesis
CLARIDGE: High-Resolution NMR Techniques in Organic Chemistry
CLAYDEN: Organolithiums: Selectivity for Synthesis
FINET: Ligand Coupling Reactions with Heteroatomic Compounds
GAWLEY & AUBÉ: Principles of Assymmetric Synthesis
HASSNER & STUMER: Organic Syntheses Based on Name Reactions
and Unnamed Reactions (2nd Edition)
LI & GRIBBLE: Palladium in Heterocyclic Chemistry
McKILLOP: Advanced Problems in Organic Reaction Mechanisms
OBRECHT & VILLALGORDO: Solid-Supported Combinatorial and
Parallel Synthesis of Small-Molecular-Weight Compound Libraries
PERLMUTTER: Conjugate Addition Reactions in Organic Synthesis
PIETRA: Biodiversity and Natural Product Diversity
SESSLER & WEGHORN: Expanded, Contracted & Isomeric Porphyrins
TANG & LEVY: Chemistry of *C*-Glycosides
WONG & WHITESIDES: Enzymes in Synthetic Organic Chemistry

Studies in Natural Products Chemistry *(series)*
Strategies and Tactics in Organic Synthesis *(series)*
Rodd's Chemistry of Carbon Compounds *(series)*
Best Synthetic Methods *(series)*

JOURNALS

Bioorganic & Medicinal Chemistry
Bioorganic & Medicinal Chemistry Letters
Tetrahedron
Tetrahedron Letters
Tetrahedron: Assymmetry

Full details of all Elsevier publications are available on www.elsevier.com or from your nearest Elsevier office

Molecular Diversity and Combinatorial Chemistry

Principles and Applications

MICHAEL C. PIRRUNG

Department of Chemistry
Duke University
Durham, NC, USA

ELSEVIER

2004

Amsterdam - Boston - Heidelberg - London - New York - Oxford - Paris
San Diego - San Francisco - Singapore - Sydney - Tokyo

ELSEVIER B.V.
Sara Burgerhartstraat 25
P.O. Box 211, 1000 AE
Amsterdam, The Netherlands

ELSEVIER Inc.
525 B Street
Suite 1900, San Diego
CA 92101-4495, USA

ELSEVIER Ltd
The Boulevard
Langford Lane, Kidlington,
Oxford OX5 1GB, UK

ELSEVIER
84 Theobal
London WC
UK

First edition 2004

Library of Congress Cataloging in Publication Data
A catalog record is available from the Library of Congress.

British Library Cataloguing in Publication Data
A catalogue record is available from the British Library.

ISBN: 0-08-044493-8 (hardbound)
ISBN: 0-08-044532-2 (paperback)
ISSN: 1460-1567 (Series)

Printed and bound by CPI Group (UK) Ltd, Croydon, CR0 4YY

Transferred to Digital Print 2012

Dedicated to my family, especially my wife Laura, and my chemistry teachers, especially Clayton Heathcock.

Table of Contents

Nature: The Original Combinatorial Chemist   1

Biopolymers constitute natural libraries   1
Selection and evolution   4
The expression of genetic information   7
Combinatorial assembly of antibody genes   10
Molecular solutions to combinatorial problems   11

Synthetic Peptide Libraries   15

Solid-phase peptide synthesis   15
Peptides on pins   17
Other iterative deconvolution strategies   20
Examples of split/couple/mix peptide libraries   24
Positional scanning   30
Epilogue   32

Supports, Linkers, and Reagents for Peptide Synthesis   37

Polystyrenes   37
PEG-grafted supports   41
Coupling strategies   42

Supports and Linkers for Small Molecule Synthesis   45

New resins and linkers   45
Ring-forming cleavage   48
Loading   49

Encoded Combinatorial Chemistry   51

Directed Sorting   59

Unnatural Oligomers for Library Synthesis   65

Peptoids   65
Azatides   66
Peptidyl phosphonates   67
Oligoureas   69

Analytical Methods for Solid-phase Synthesis   71

Product identification   71
Gel-phase NMR   71
High-resolution magic angle spinning NMR   72
On-bead infrared spectroscopy   73
Mass spectrometry   74
Non-spectroscopic methods   75

Supported Solution-phase Synthesis   79

Polyethylene glycols   79
Dendrimers   80
Fluorous synthesis   80

Solution-phase Parallel Synthesis   85

Scavenging resins   85
Ion-exchange resins   86
Supported reagents   88
Fluorous reagents   90
Solid-phase extraction   91
Gas-phase separation   91

Multi-component Reactions   93

Chemical Informatics, Diversity, and Library Design   101

Strategies   102
Representative flowchart for a library design   103

Lipinski's Rules   105

Nucleic Acid Microarrays   107

Combinatorial Materials Chemistry   115

Combinatorial Catalyst Discovery   121

Peptides on Phage   131
Applications of phage display   136

Nucleic Acid Selection   141

Complex Combinatorial and Solid-phase Synthesis   149
Epibatidine   149
Mappicine   150
Dysidiolide   152

The Big Picture   155

Answers to exercises   157

Bibliography   167
Major research texts on combinatorial chemistry   167
URLography   168
Reviews   169

Index   171

Foreword

Molecular diversity and combinatorial chemistry represent relatively new approaches to solving problems in chemical reactivity and function. Traditional approaches to gain substances with desirable new function focused on understanding the function of existing systems and establishing the principles governing their performance, permitting the rational design of new systems with improved performance. While undoubtedly effective, as two centuries of chemistry attests, this approach is laborious and time-consuming. An alternative approach has been called "blind watchmaker" (after R. Dawkins's book of that title) or "Edisonian" (based on the report that Thomas Edison randomly tried >6,000 different "vegetable growths" as incandescent filaments before discovering carbonized cotton thread). It involves merely trying things until you find one that works well. While unappealing to many scientists for its lack of intellectual content, this approach can be very effective. It is even more effective if the trials in this trial-and-error method can be made very fast and inexpensive. The intellectual challenge for modern scientists in using such methods, then, is to develop techniques that permit many trials to be performed in a parallel or rapid, serial way. Choice of the types of chemical inputs to such techniques is also crucially dependent on chemical knowledge, rationality, and judgment.

While there are many scientific treatises on combinatorial chemistry, and many of them incorporate the diversity that is representative of such a rapidly developing field, most are aimed at the scholarly practitioner, not the aspiring scholar. This textbook is an effort toward making the main concepts of the field accessible to non-specialists. It is written with the advanced undergraduate student or graduate student in mind and would be appropriate for a one-quarter advanced topics course.

Preface

This book had quite a long gestation. A generous 1997 invitation from Chris Leumann to lecture in the "Troiseme Cycle en Chimie" of universities in the French-speaking regions of Switzerland resulted in materials for a combinatorial chemistry short course. We had a wonderful week of fellowship and science high in the mountains in Champéry. A second, major impetus was an opportunity offered by Peter Dervan to teach a combinatorial chemistry course as a visiting professor at Caltech in 2002. The lecture notes emerging from that scientifically stimulating quarter in Pasadena provided much of the current form of the book. Caltech also provided a reminder of the classic short textbook "Organic Synthesis" by Bob Ireland, published over three decades ago, providing further inspiration to develop this manuscript. Subsequent classes at Duke and UC-Irvine refined the material, and the input and perspective from reviewers and editors enabled its 2004 publication.

Likewise, combinatorial chemistry has had a long gestation. Historians of science may be able to trace its conception more carefully than I, but the subject reached critical mass in the early 1990s, with important initial publications emerging from many labs around that time. The cliché "Success has a thousand fathers, but failure is an orphan" makes the identification of a true father of the field a foolhardy pastime, but a "baby boom" in combinatorial chemistry certainly occurred around that time. The field has become well established over the past decade. While not a panacea for all chemical problems, combinatorial chemistry is clearly very powerful when applied to problems appropriate to its principles. It has become one of the many important strategies for chemical problem-solving.

Acknowledgments

Nick Paras and Jim Falsey, then graduate students at Caltech, were valuable contributors to my first efforts to develop a full-blown combinatorial chemistry course, and did yeoman service in preparing problem sets that were developed into the exercises in this book. Students who listened to my lectures, studied my notes, and evaluated my courses made contributions that they may not recognize but are nonetheless appreciated. Deirdre Clark and Ian Salusbury at Elsevier were very generous with their time and advice, both literary and technical. My administrators over several years, Laurie LaBean and John Wessels, tracked down many an obscure reference and facilitated this work in many other ways. I appreciate the provision of original figures by James Morken (North Carolina), Dennis Curran (Pitt), Wei Zhang (Fluorous Technologies), Donald Burke (Indiana), Elizabeth Fretz and Barry Prom (Discovery Partners International) and Alireza Givehchi (CallistoGen). Finally, members of my laboratory did their customary excellent job proofreading the manuscript.

Michael C. Pirrung

Chapel Hill, North Carolina
April, 2004

Nature: The Original Combinatorial Chemist

Biopolymers constitute natural libraries

Nature must create a wide variety of functional molecules to support life, and it does this from only a few basic starting materials using combinatorial principles. Two of the major classes of biopolymers, proteins and nucleic acids, are inherently combinatorial and are constituted from amino acids and nucleotides, respectively. We can call these sets of molecules *building blocks*.

The structures of the four deoxynucleotides that make up DNA are given below, with their one letter codes and identification as purines (the bicyclic nitrogen heterocycles) or pyrimidines (the monocyclic nitrogen heterocycles) based on the structure of the heterocyclic base.

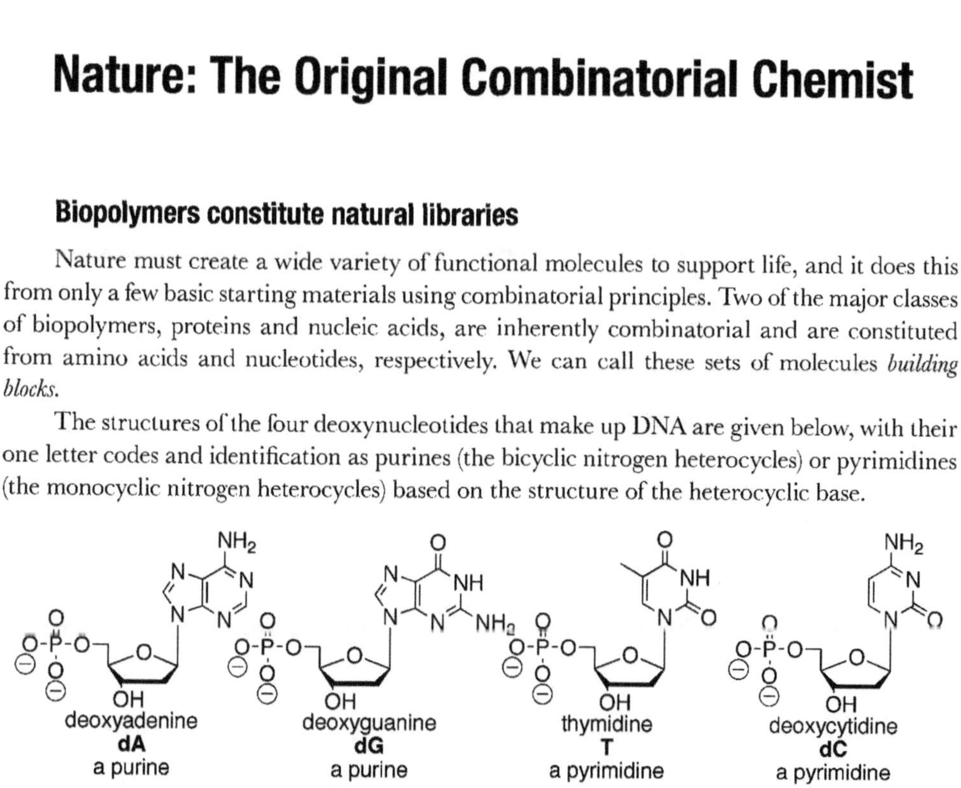

deoxyadenine	deoxyguanine	thymidine	deoxycytidine
dA	**dG**	**T**	**dC**
a purine	a purine	a pyrimidine	a pyrimidine

The four deoxynucleotides

The structures of the four nucleotides that make up RNA are given below. They are distinguished from the deoxynucleotides by their ribose sugar and the absence of the methyl group of thymidine, making uridine.

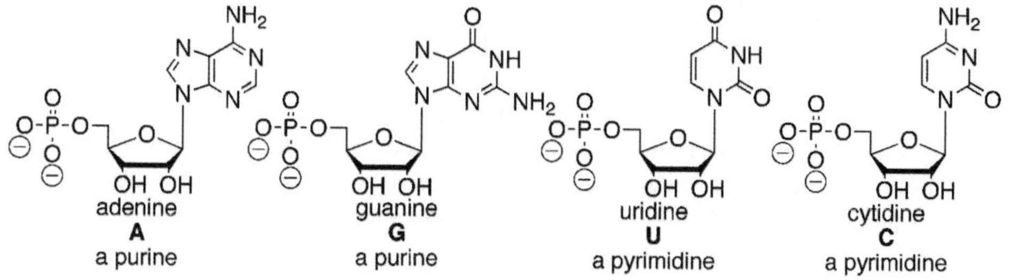

adenine	guanine	uridine	cytidine
A	**G**	**U**	**C**
a purine	a purine	a pyrimidine	a pyrimidine

The four nucleotides

The structures of the twenty genetically encoded amino acids are given below, with their three-letter and one-letter amino acid codes.

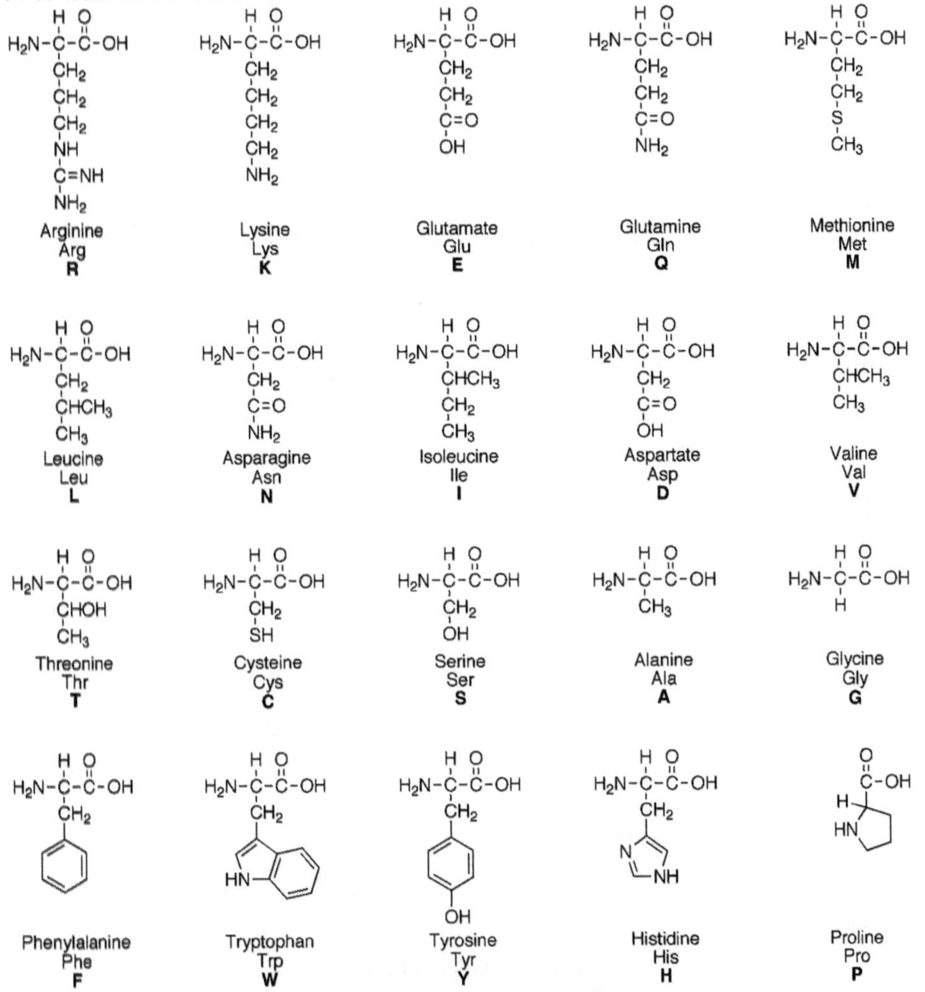

Arginine
Arg
R

Lysine
Lys
K

Glutamate
Glu
E

Glutamine
Gln
Q

Methionine
Met
M

Leucine
Leu
L

Asparagine
Asn
N

Isoleucine
Ile
I

Aspartate
Asp
D

Valine
Val
V

Threonine
Thr
T

Cysteine
Cys
C

Serine
Ser
S

Alanine
Ala
A

Glycine
Gly
G

Phenylalanine
Phe
F

Tryptophan
Trp
W

Tyrosine
Tyr
Y

Histidine
His
H

Proline
Pro
P

The twenty amino acids

Each member in a set of these building blocks has a pair of complementary functional groups. For example, the hydroxyl group of deoxyadenine can be linked to the phosphate of deoxyguanine to form a phosphate ester and create a dinucleotide. The amino group of glycine can be linked to the carboxyl group of alanine to form an amide and create a dipeptide. These building blocks have one of each type of functional group, creating unsymmetrical attachments to other building blocks and making the oligomers created from them directional. Thus, linking of alanine and glycine can be done in two different ways, one in which the glycine carboxylate is used to form the amide, and one in which the alanine carboxylate is used. The molecules Ala-Gly (top, following page) and Gly-Ala (bottom, following page) are different.

Formation of nucleotide and amino acid dimers

Molecules are built up by incorporating a specific number of units drawn from the building block set, defining the length of the molecule. Each position in the chain along its length is unique because of the directionality of these polymers. The directionality is designated by the nomenclature of the individual building blocks. Thus, the two different ends of (poly)nucleic acids can be designated 3' or 5', and the two different ends of (poly)peptides can be designated amino (N-) terminal and carboxy (C-) terminal.

The product of each linking process possesses the same functional groups as the individual building blocks, enabling linking to be repeated recursively to build up oligomers. We now consider how many *different* molecules are created when a specified set of building blocks is linked together in all combinations. Because of the directionality of the oligomers, both the order of the building blocks (their *sequence*) and their identity are important. Using the convention that an arrow signifies a directional bond, this means that:

$$\text{A}\rightarrow\text{B}\rightarrow\text{C}\rightarrow\text{D} \neq \text{D}\rightarrow\text{C}\rightarrow\text{B}\rightarrow\text{A}$$

where **A** through **D** represent any of the building blocks for the particular oligomer being constructed. If the bonding scheme were not directional, then

$$\text{A-B-C-D} = \text{D-C-B-A}$$

Directional oligomers are thus somewhat analogous to non-commutative operations in mathematics, while non-directional oligomers are analogous to commutative operations.

By convention, peptide and protein sequences are written with the amino acid that has the free amino group on the left, and with the amino acid that has the free carboxyl group on the right. DNA sequences are often written with the 5'-end on the left and the 3'-end on the right. Strong adherence to these rules is not universal, so designation of the ends is important. The complementary strand of DNA need not be explicitly written since the base-pairing rules define its sequence. However, it is possible to form double helical DNA with non Watson-Crick base pairs at some positions. In these instances, it is necessary to write both strands. If both strands are written, the "top" strand often has its 5'-end on the left and the "bottom" strand has its 5'-end on the right. This directional switch is a direct consequence of the fact that nucleic acids form anti-parallel duplexes.

5'-AGTCAGTTA-3'
3'-TCAGTCAAT-5'

For a building block set of **N** members, any building block can be used at any position, giving **N** possibilities at each. The number of possible sequences is the product of the number of possibilities at each position in the sequence, or:

$$N \times N \times N \times N = N^4 \text{ (in this instance).}$$

More generally, the number of possibilities is an exponential function of the number of building blocks and the length of the oligomer:

$$N^l, \text{ where}$$

$$N = \# \text{ of building blocks and } l = \# \text{ of positions.}$$

For the nucleic acids, **N** is 4, so the number of possible DNA or RNA sequences of length 10 is 4^{10} or about 1 million. It is estimated there are 3.5 billion nucleotides within the human genome, making the number of possible sequences $4^{3,500,000,000}$, truly beyond conception. On the other hand, the human genome is a specific sequence, and the probability of finding any particular subsequence within it is calculable. With the useful (but incorrect) assumption that human DNA is totally random, a given sequence of only 17 nucleotides is predicted statistically to match few local sequence sites within the whole human genome, because 4^{17} is greater than 3.5 billion. Allowing for DNA to be non-random, as it must be to perform the functions of life (there may be repetition of sequences with important functions), it is still expected that most individual sites within the human genome can be uniquely identified with sequences of 25 or so nucleotides, because 4^{25} is vastly larger than 3.5 billion. The probability of a site being unique increases with the length of the sequence. Unique sequences that constitute mileposts along the human genome have been catalogued in the course of determining the structure of human DNA (specifically, sequencing the DNA within the genome that encodes proteins). Such unique sites are called *expressed sequence tags*, or ESTs. The "expressed" refers to the fact that the DNA sequence is converted into protein using the principles described in a following section. For genetically coded amino acid oligomers (peptides and proteins), **N** is 20. Therefore the number of possible pentapeptides composed of the genetically coded amino acids is 20^5 or 3.2 million. The division between peptides and proteins is ill-defined, but proteins would certainly include those polypeptides with greater than 100 residues. Using 100 as a lower limit, there are at least 20^{100} possible proteins; of course proteins can be much longer, up to thousands of amino acids, making the conceivable diversity of polypeptides truly astronomical, greater than the number of atoms in the Universe.

Selection and evolution

Clearly, the sequences of the proteins that are the engines of life are not randomly selected from among the vast number of possibilities. These proteins must perform specific functions, and the only mechanism acknowledged for achieving function in Nature is natural selection, or evolution. Evolution is most often considered as operating at the level of the organism, with

more fit individuals reproducing more successfully, and less fit individuals reproducing less successfully. It may also operate at the level of the sequences of genes and the proteins they encode,
since it is these molecules that underlie the function (and therefore fitness) of the organism.

Consider a thought experiment beginning with the 400 (20^2) genetically encoded dipeptides. If we could examine each of these for a desired function, say catalysis of the hydration of
carbon dioxide, the single sequence that performed the reaction best could be allowed to survive, and the other 399 terminated. This would constitute one "generation" of molecules. That
one best catalyst could be used as a core sequence in 40 tripeptides (built up by adding amino
acids at both the C-terminus and N-terminus) that would again be tested for the desired function, constituting the second generation. Again, the one best tripeptide catalyst would be used
as a core sequence to build up 40 tetrapeptides. This recursive process could continue to build
up a protein that would hopefully have sufficient activity to perform the function required. This
is a sort of evolution at the molecular level. While Nature surely did not use this procedure to
evolve the sequence of the enzyme carbonic anhydrase that catalyzes the hydration of carbon
dioxide (most specifically, because carbonic anhydrase uses a metal ion in its catalysis), some-
thing like this could have applied to primordial life forms that were improving their ability to
survive and replicate by random or more directed modification of the functions of the molecules
within them.

Recursive synthesis/testing is evolution at the molecular level

One of the key advantages to the operation of this recursive process of serial improvement
in Nature is that organisms have the capability of self-replication, and consequently they are
able to replicate all of the molecules within themselves. In the chemical world, compounds must
generally be prepared time and again by synthetic procedures, though in select cases molecules
have the capability for self-replication, DNA being the best example.

Natural selection and self-replication provide powerful tools to the scientist wishing to
engineer living systems. We will use as an example the process of bacterial transformation, a
widely practiced laboratory procedure in molecular biology. Broadly defined, transformation
is simply the taking up of foreign DNA by a bacterial cell. This capability can be harnessed by
the genetic engineer in order to place foreign DNA into bacterial cells so it can be *expressed* as
a protein. One way foreign DNA can be carried into a cell is via a *vector* called a plasmid. This
process occurs naturally, and is one means that traits like antibiotic resistance are spread in a
bacterial population.

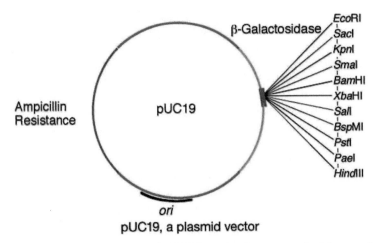

pUC19, a plasmid vector

A *plasmid* is a circular, double-stranded DNA molecule composed of thousands of nucleo-tides. A plasmid commonly used in the molecular biology laboratory is pUC19. DNA can be inserted into a site in the β-galactosidase gene shown at the right. The abbreviations given there identify many possible enzymes that can cut the plasmid to permit the insertion of foreign DNA. On the left in blue is the gene for ampicillin resistance that encodes the β-lactamase protein; it degrades β-lactam antibiotics. The plasmid contains one other essential function: an origin of replication (*ori*). When the cell is preparing to divide, this specific sequence of DNA is a signal that tells the cell to copy the DNA of the plasmid at the same time as the overall DNA of the cell is being copied. This ensures that both daughter cells will contain the same plasmid as the mother cell.

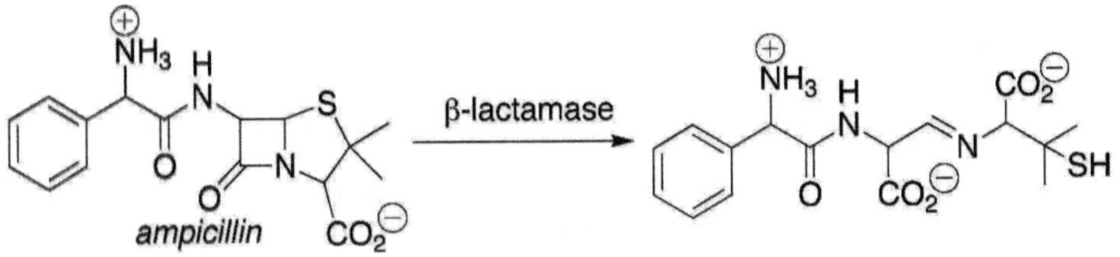

β-Lactamase hydrolyzes ampicillin

The pUC19 vector is used in transformation as follows. Under special conditions, such as high concentrations of calcium ion and/or temperature shock, *E. coli* cells will take up the plas-mid. Of course, not every cell will do so, so we are faced with the issue of separating those cells that have been transformed from those that have not. This would be a very challenging process to address chemically, but it is trivial for biology. When grown in the presence of the antibiotic ampicillin, those bacteria that do not contain the plasmid will be killed. Those that have taken it up will be resistant because they can hydrolyze the antibiotic to an inactive compound. They will grow and can come to dominate the culture. When the dilute culture is poured out onto a petri plate containing growth media including the antibiotic, small bacterial colonies will grow up that are derived from single progenitor cells. Since the two daughter cells resulting from mitosis are identical to the mother cell, all of the cells in a single colony will be genetically identical. They are therefore *clones*, and all will contain the plasmid.

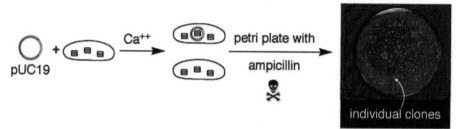

Transformation and selection of transformants

To summarize, two very powerful characteristics of biological systems, self-replication and natural selection, are crucial for the true evolution of function. Self-replication is exemplified by cell division. Since each mitosis event makes two identical copies of a mother cell, 30 cell divisions will lead to 2^{30} (over 1 billion) genetically identical daughter cells. Natural selection permits a single cell with a desired trait to be selectively replicated even when it is in a large population of other cells. Through the natural amplification arising from self-replication, that cell can come to dominate the population to the exclusion of others.

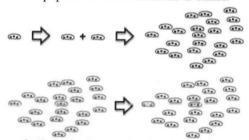

Self-replication and natural selection

The expression of genetic information

Combinatorial principles are commonly seen in Nature, one example being the genetic code. The DNA in the genome of an organism provides the instructions for the production of a specific protein molecule via the intermediate molecule RNA (actually, messenger RNA, or mRNA). Thus, the central dogma of molecular biology is:

DNA makes RNA makes protein

and the corollary of the central dogma is "one gene, one protein." The process of converting DNA to RNA is called transcription, and the process of converting RNA to protein is called translation. Because the DNA and RNA molecules are fairly similar, transcription is analogous to changing a font, while the RNA and protein molecules are very different, so translation is analogous to changing the language.

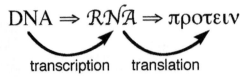

transcription   translation

The central dogma of molecular biology

To further understand how the central dogma plays out in practice, we will describe the production of a single protein. One interesting feature that immediately arises that exemplifies the combinatorial principles of protein production is that there are 20 amino acids but only 4 nucleotides to encode them. As even novice combinatorial chemists could predict by now, a means to deal with this problem is simply to lengthen the nucleotide molecule that encodes the amino acid. Going from a single nucleotide to two linked nucleotides increases the possibilities for nucleotide sequence from 4 to 16. Sixteen nucleotide combinations are still insufficient to encode 20 amino acids, however, so a third nucleotide is required. The use of trinucleotides (also called triplet codons) results in 64 possible nucleotide combinations, more than enough for the set of amino acids. The encoding of amino acids by three nucleotides has two ramifications. One is that the genetic code is degenerate, in that different combinations of nucleotides can correspond to the same amino acid. The other is that some of the combinations are available for other functions. Three of these combinations are used to signify that the protein ends. One combination, AUG, always starts the protein and encodes the amino acid methionine. All proteins thus begin with methionine at their N-terminus.

The sequence of three nucleotides that corresponds to a particular amino acid is called a *codon*. The 64 possible codons and the amino acids they encode (or stop codons) are summarized in the table on the facing page. The transcription and translation of an example DNA sequence is shown below. The transcription of the DNA sequence leads to an RNA molecule that is essentially identical in sequence and very similar in chemical structure. The ribose sugar replaces the deoxyribose sugar, and U is substituted for T. This is only a slight modification, since U has the same preference to form base pairs with A as does T. When the mRNA is translated into protein, the nucleotides are read as codons, or three at a time. There is no specific marker for codons, however, a phenomenon that was called "the comma-less code" by Francis Crick. The analogy is to writing, where items are separated by commas so as to distinguish "eats, shoots and leaves" from "eats shoots and leaves." Because codons are not specifically marked, it is crucial that the enzymes involved in replicating and transcribing DNA not make even a single nucleotide addition or deletion. Such a mistake would disrupt the *reading frame* of the RNA and lead to the production of a wholly different protein than that originally encoded.

in DNA   5'-ATG GAA CAA GTA GGA-3'

in mRNA   5'-AUG GAA CAA GUA GGA-3'

protein H_2N-Met Glu Gln Val Gly-CO_2H

Correspondence between DNA, RNA, and protein

Table 1. The genetic code

Second codon position

		U	C	A	G		
	U	UUU } Phe UUC } UUA } Leu UUG }	UCU } UCC } UCA } Ser UCG }	UAU } Tyr UAC } UAA } STOP UAG }	UGU } Cys UGC } UGA STOP UGG Trp	U C A G	
	C	CUU } CUC } Leu CUA } CUG }	CCU } CCC } CCA } Pro CCG }	CAU } His CAC } CAA } Gln CAG }	CGU } CGC } CGA } Arg CGG }	U C A G	
	A	AUU } AUC } Ile AUA } AUG Met	ACU } ACC } ACA } Thr ACG }	AAU } Asn AAC } AAA } Lys AAG }	AGU } Ser AGC } AGA } Arg AGG }	U C A G	
	G	GUU } GUC } Val GUA } GUG }	GCU } GCC } GCA } Ala GCG }	GAU } Asp GAC } GAA } Glu GAG }	GGU } GGC } GGA } Gly GGG }	U C A G	

First codon position (left axis) — Third codon position (right axis)

At the DNA level, a gene is more than just the sequence of codons that corresponds to the protein sequence (the structural gene), it includes flanking regions that provide signals for the complex process of gene expression. For example, the promoter has a binding site (a TATA sequence) for the RNA polymerase that performs transcription to create the messenger RNA. A sequence in the 5'-untranslated region of the mRNA is the recognition/binding site for the ribosome, the RNA-protein complex that actually performs protein synthesis.

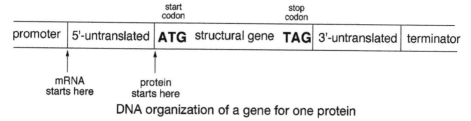

DNA organization of a gene for one protein

As discussed earlier, the length of the protein can be hundreds to thousands of amino acid units. The length of the structural gene must be 3× the length of the protein, but the length of the overall DNA that constitutes the gene for the protein (including flanking control sequences) could be hundreds to thousands more nucleotides.

Combinatorial assembly of antibody genes

A well-known application of combinatorial principles in Nature is in the assembly of the protein structure of antibodies. The vertebrate immune system can create molecules that bind to essentially any type of compound, and it does this through the amazing diversity of antibodies. It is estimated that 100 million different antibodies are represented within our genomes (the primary repertoire). Even more amazing is that these 100 million protein molecules do not arise from 100 million genes. This situation bends (but does not break, as we will see) the "one gene, one protein" rule.

The overall topology of one of the simplest types of antibodies, the immunoglobulin G type, is shown below. The IgG class is assembled from two copies each of a heavy chain and a light chain (larger and smaller protein molecules, respectively). The carboxyl-terminus of each chain is indicated with a C. Yellow dashed lines indicate disulfide bonds that form loops along the protein chain and hold different chains together. The V_L and V_H loops are the variable regions contributed from the light and heavy chains that constitute the binding site of the antibody. The IgG molecule has two identical binding sites.

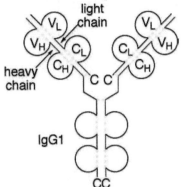

The structure of IgG1

Variation in the protein structure is achieved by having a number of possibilities for each of the functional segments (called the V, D, and J regions, for variable, diversity, and joining) of the gene for the antibody. Any one of these segments, but only one of these segments, can be used in the creation of the gene that encodes the final antibody molecule. In the human IgG heavy chain DNA sequence, beginning at the 5' end, there are about 200 possible V_H segments. These are followed by about 30 D_H segments and then 6 possible J_H segments. In the DNA sequence of one particular IgG1 molecule, the κ type, the light chain sequence begins with about 100 possible $V_κ$ segments, followed by 5 possible $J_κ$ segments.

This is the arrangement in the genome as a whole and in naïve immune cells that have not been exposed to antigen (a molecule that elicits an immune response). Each mature immune cell produces only one specific antibody molecule from a specific combination of V, D, and J gene

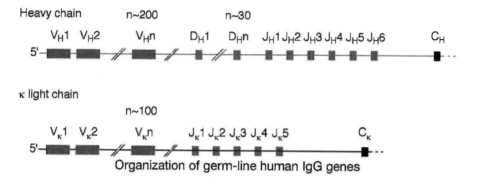

Organization of germ-line human IgG genes

segments. This is so because genetic recombination rearranges the germ-line DNA sequence, with the result that in the final heavy chain gene that is expressed within a single immune cell, only one V segment is joined to only one D segment, which is joined to only one J segment. In the final light chain gene, only one V segment is joined to one J segment. In the example given below, the heavy chain would be composed from the V_H1, D_H1, and J_H4 segments, and the light chain from the $V_\kappa2$ and $J_\kappa2$ segments.

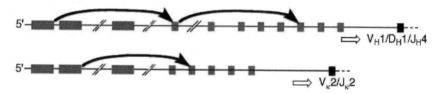

Recombination of IgG gene segments

With this knowledge, we can calculate the number of possible arrangements of human IgG1κ genes. Presuming 100 V and 5 J sequences for the light chain and 200 V, 30 D, and 6 J sequences for the heavy chain, after recombination there will be

$$100 \times 5 \times 200 \times 30 \times 6 = 1.8 \times 10^7,$$

or over 10 million different possibilities. With several types of antibodies other than IgG in the vertebrate immune system, the overall diversity of antibody molecules is estimated to be $> 10^8$. Even greater diversity is introduced by mutations in the DNA that occur during maturation of the immune response. This vast diversity underlies the widely discussed idea that the immune system can create antibodies to bind any molecule.

Molecular solutions to combinatorial problems

A particularly difficult class of problems in computation are the so-called NP-complete problems, where "NP" stands for non-deterministic polynomial time. The time required to solve NP-complete problems increases exponentially with the number of variables. They can be trivial to solve with one or two variables, and impossible to solve within human lifetimes with ten or twenty variables. NP-complete functions are used in areas such as data encryption, where the current standard is 128-bit encryption (2^{128} combinations). NP-complete problems are also called combinatorial problems, and the analogy is accurate — combinatorial molecules increase in size exponentially with the number of variables (positions), as we have seen. Combinatorial principles from chemistry thus might be applied to code-breaking, for example. Molecular computations using DNA have been developed with an eye toward this goal.

Another relation between NP-complete problems and Nature can be perceived. Evolution is the only acknowledged mechanism to create function that is available to Nature. Variants (available through natural genetic diversity, mutation, and sexual reproduction) are challenged for fitness in a task, and those with superior performance survive and proliferate into the next generation; those that are less fit are overcome. This approach will succeed best with the largest number of variants as the starting point. Hence it is good (and/or lucky) that biological molecules such as nucleic acids/proteins are combinatorial (NP-complete).

One can envision forcing molecules to undergo evolution in the laboratory, if they have the capability of self-replication. As we will see later, this has been accomplished to select nucleic acids with functions, such as binding particular ligands. However, most molecules that chemists want to discover will not be self-replicating. This eliminates the earlier-described method, selection and amplification, to obtain a high functioning molecule. It also means that all of the molecules to be tested must be synthesized and then screened for the desired property. Molecules may then be chosen to serve as the basis for another generation of synthesis and screening.

Given this paradigm, efficient methods to synthesize many variants of molecules are needed; efficient methods to screen them are equally important. Replacing serial processes in synthesis and screening that are traditional in chemistry with parallel processing will greatly facilitate achievement of this goal.

Close integration of both the synthesis and screening phases of the combinatorial process is necessary, and in fact the argument can be made that the synthesis of a molecular library without some type of testing is an incomplete experiment. Furthermore, the types of testing that will be applied to a library must be incorporated into the synthetic planning. The assay influences the form of the library (pure compounds, pools of compounds, self-replicating compounds, non-self-replicating compounds) as much as the synthesis method. A powerful method for creation of a combinatorial library would be useless if the library could not be screened in the assay that was intended. The most common circumstance in which this issue must be addressed is the screening of compounds while bound to a solid phase or in solution.

For the most part, the tasks that have been used to test combinatorial principles have been in pharmaceutical development, though general organic molecular recognition, catalyst discovery, superconductors, and vaccines have also been pursued.

Additional reading

"Recombinant DNA" - Watson, J. D.; Gilman, M.; Witkowski, J.; Zoller, M. Scientific American/W. H. Freeman, NY, 1992.

Problems

1. The gene for the cell surface molecule CD44 is composed of 20 different exons, the gene segments that are spliced together to make mRNA. The first 5 exons and the last 5 exons are always included, but depending on the developmental stage of the cell, any of the exons 6-15 can be included or not. How many combinations of CD44 mRNAs are there?

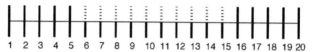

1 2 3 4 5 6 7 8 9 10 11 12 13 14 15 16 17 18 19 20

2. To see the effect of reading frame on protein sequence, insert a G following the start codon in the RNA sequence 5'-AUG GAA CAA GUA GGA-3' from the example in the text and use the genetic code table to translate the resulting sequence into protein.

3. How many different DNA sequences of 60 nucleotides in length are there? How many different peptide sequences can be encoded in DNA of 60 nucleotides in length?

4. In living systems, what two traits are essential for autonomous improvement in function? Describe an example of each.

5. Calculate the number of oligomers four units in length that can be composed from these α-hydroxy acids.

Synthetic Peptide Libraries

Combinatorial techniques first arose in the peptide field. This is likely because there was already significant familiarity with the combinatorial nature of peptides, and methods for peptide synthesis on solid phase were well worked out. Solid-phase synthesis is essential to some combinatorial synthesis methods, such as the very powerful split/mix technique.

Solid-phase peptide synthesis

A brief review of solid-phase peptide synthesis (SPPS) techniques is appropriate. Like most bifunctional building blocks, amino acids must be differentially protected or activated in order to allow selective bond formation to create a specific target sequence. The simultaneous presence of a free amino group and an activated carboxylic acid in the same molecule would cause uncontrolled polymerization. The amino group is thus protected with a carbamate group. Two types of carbamates are commonly used, Fmoc and t-Boc (most often called just Boc). The latter is removed with trifluoroacetic acid (TFA), and the former is removed with piperidine. The amino acids also bear other protecting groups for the functionalities in their side chains that would interfere with peptide bond coupling. These may include the carboxyl groups of glutamate (glu or E) and aspartate (asp or D), the basic nitrogens of arginine (arg or R) and histidine (his or H), and the thiol of cysteine (cys or C). In the example given on the following page, cysteine is protected as its acetoxymethyl (Acm) derivative. Amino acids protected with Boc or Fmoc on their α-amino groups and with appropriate side chain protection are commercially available. These include the 20 genetically coded amino acids as well as a wide range of unnatural amino acids, including in many cases both the (R) and (S) enantiomers.

The growing peptide chain is linked to a beaded polymeric support during synthesis. This makes the isolation of reaction products simple, involving only filtration for removal of excess reagents and byproducts. This advantage is offset by the disadvantage that there is no opportunity for purification of intermediates during a synthesis. Therefore, the products from any reaction failures contaminate the final product and must be removed once the finished peptide has been released from the bead. Like any synthesis, the yield in a SPPS is the product of the yields of the individual steps. However, without intermediate purification, the product of the individual step-wise yields also determines the purity of the crude peptide. Since purification of a desired peptide from a number of closely related peptide sequences can be quite challenging, even with modern chromatographic methods, in SPPS a premium is placed on highly efficient reaction chemistries. Ideally, the step-wise coupling yields should be 100% and the peptide will be pure when removed from the support. Because of the importance of peptides in biology, which has motivated the concerted efforts of a significant cohort of chemists over several decades, SPPS can often approach this ideal. This iterative process is amenable to automation, and peptide synthesis machines are available commercially.

Synthesis begins with linking of the first amino acid to the support. Supports for solid-phase synthesis will be discussed in the following section. Each support bead bears many copies of the functional group that is used to attach the amino acid. The number of groups is naturally a function of the size of the bead. For example, a 90 micrometer (μm) diameter TentaGel® bead bears 6×10^{13} functional groups (100 pmol), while a 130 micrometer diameter bead bears 3× that. A common support is polystyrene that has been functionalized with chloromethyl groups (Merrifield resin). The cesium salt of a Boc-protected amino acid (cysteine in the example) displaces the chloride, linking the amino acid to the support through its carboxyl group. The *N*-protecting group is removed, making its amino group available for reaction.

Principles of solid-phase peptide synthesis

Another amino acid that is protected at its α-amino group (in the example, histidine protected with a toluenesulfonyl group on its heterocyclic nitrogen) is activated for coupling at its carboxyl group. A very wide variety of reagents can be used to activate the *N*-protected amino acid, including anhydrides of carboxylic and phosphoric acids, carbodiimides, and uronium salts. A common feature of these activation methods is that they generate a good leaving group for an acyl substitution reaction, most commonly by forming a stabilized anion. This activation step offers the greatest opportunity for racemization of the stereogenic center adjacent to the carboxyl group. The specific activating reagent and other additives such as *N*-hydroxybenzotriazole (HOBt) can minimize the amount of racemization in this step.

The activated amino acid reacts with the amino group on the support to create the first peptide bond. This reaction may be incomplete, especially with amino acids bearing bulky groups, creating a resin that has some but not all of the specified amino acid coupled at this position. If synthesis were to continue from this point without further defects, two peptides would be produced, one of full length and one with a one amino acid deletion from the sequence. It is thus important to ensure that no unreacted amino groups remain after coupling. The extent of coupling can be determined by testing the support beads for free amino groups (as contrasted with those within peptide and carbamate bonds) using colorimetric reagents such as bromophenol blue or ninhydrin. If coupling is complete, no amino groups should be detected. When complete coupling is not observed, it can be repeated with the same amino acid (*double coupling*), or the support can be treated with a very reactive acylating agent such as acetic anhydride (*capping*). Capping converts uncoupled amino groups to their acetamides and prevents them from being extended in further cycles of synthesis. Capping does not affect the amount of byproducts in the final crude peptide, but they differ in length from the desired target sequence (they are truncation sequences) and can be more easily separated from it than the alternative deletion peptides would be.

Iteration of the "deprotect" and "couple" steps constitutes a synthesis cycle that extends the peptide by one amino acid. The cycle yield can be determined as the synthesis proceeds to ensure the purity of the final product. For example, the byproduct of deprotection of the Fmoc group is (fluorenylmethyl)piperidine, which has its strongest UV absorption (ε 17,500) at 267 nm. Its concentration can be directly determined by UV spectroscopy and the application of Beer's Law. The concentration can be related back to the amount of Fmoc-protected amino acid that was coupled in the previous step and subsequently deprotected.

Once all of the amino acids are incorporated, the protecting groups on the side chains must be removed. Most commonly, they are labile to acidic conditions. For Boc chemistry, which uses acid for deprotection during peptide chain assembly, the side-chain protecting groups must be stable to these conditions, so stronger acid is needed for side-chain deprotection. Acid is also used to cleave the ester bond to the simple Merrifield resin by an Sn1-like mechanism. Both steps can be accomplished with liquid HF. The hazards and inconvenience of working with HF (high pressure gas manifolds, the inability to use glassware, and a highly poisonous reagent) prompted the development of the alternative Fmoc chemistry. Since deprotections during the coupling cycles are performed with a base in Fmoc chemistry, the side chain protection need not be as acid-stable as in Boc chemistry. Commonly, Boc groups and *tert*-butyl esters are used to protect amino and carboxy groups in the amino acid side chains in Fmoc chemistry. Thus, at the conclusion of the synthesis they can be removed with TFA. The linkage of the peptide to the resin is also more acid labile than in Boc chemistry.

Peptides on pins

The chemistry of peptide synthesis as described above is not dependent on the identity of the support, permitting Geysen to develop a method to perform peptide synthesis on polyethylene pins arrayed in the 8 × 12 format of the 96-well microtiter plate. These plates are commonly used in biological assays, and are readily adapted for chemistry. Different reagents for coupling different amino acids can be placed into the wells by simple pipetters; automated equipment for both rapid solvent addition and removal based on this format is also widely available. The array of pins can be simultaneously dipped into the wells of reagents to perform the coupling steps on each. Thus, 96 peptide coupling reactions are conducted in parallel. Furthermore, the sequence of the peptide on each pin would be known based on the sequences of reagents to which the pins at each position were exposed. This was the first example of a spatially defined library of compounds, where the molecular identity could be known based on physical position.

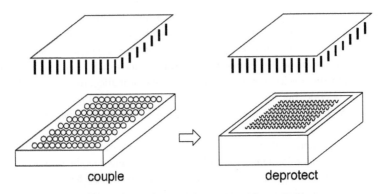

couple   deprotect

Peptide synthesis on pins using 96-well plates

The peptides synthesized with this method are not intended to be removed from the pin, so any type of functionalization of the plastic (linker) that terminated in an amino group could be used. Acrylic acid polymers can be radiation grafted onto the pins, a diamine linker coupled to the resulting carboxyl groups, and two β-alanine residues added to the diamine. The linker terminates in an Fmoc-protected amine. The pins are pressed into holes in a plastic block in the same 8 × 12 arrangement as the microtiter plate. Fmoc peptide synthesis is performed using the phosphonium salt BOP (benzotriazol-1-yloxy)tris(dimethylamino)-phosphonium hexa-fluorophosphate) as the coupling agent.

Linker and coupling agent for pin synthesis

After coupling, the whole block of pins can be treated with piperidine in batch mode to remove the *N*-terminal Fmoc protecting groups. This cycle is repeated until peptides of the desired length are prepared. The *N*-terminal residue is usually acetylated, and the protecting groups on the amino acid side chains are cleaved by trifluoroacetic acid. Heptapeptide libraries can be prepared in only three days using this method.

The problem that Geysen initially aimed to address with this method for peptide synthesis was mapping of epitopes. An *epitope* is simply the region of a protein that is recognized by an antibody. This is often a contiguous sequence of 4-10 amino acids. He sought a method to determine this region systematically and directly by simply preparing all of the n-mers within the protein sequence and testing them for reaction with an antibody. For example, hexapeptides would be prepared such that consecutive peptides overlap by five amino acids. A hexapeptide "window" would therefore be scanned down the protein sequence, looking for a site of reaction with the antibody, so the method was named epitope scanning. Since hundreds or thousands of amino acids can be present in the sequence of a given protein, epitope scanning required a method to prepare about this number of individual peptides for testing, and peptides on pins fulfilled this need.

For a number of protein/antibody pairs, epitope scanning using peptides on pins does not identify any contiguous peptides that bind strongly. Presumably this is because the antibody does not recognize a linear epitope but a "discontinuous" epitope, one composed from amino acid residues that might be far apart in the primary sequence of the protein but close to one another in the folded protein. The synthesis of candidate peptides could therefore not be based on the protein sequence. Given that an antibody is thought to recognize about six amino acids, one solution to the problem would be to prepare all of the hexapeptides (20^6 or 64 million!) and compare their reactivity with the antibody. This number is far too large to be prepared one at a time even by the pin method.

Epitope scanning by peptide synthesis

Geysen's solution to this need for a large number of candidate peptides was to prepare pools of peptides on each pin. Some positions are defined and some positions are mixtures of all 20 amino acids. A seemingly equivalent strategy, somewhat like the thought experiment on protein evolution discussed earlier, would involve preparing and assaying short peptides to identify some residues and then extending the sequence to find the next residue outside this "kernel". This approach is not attractive because antibodies bind relatively weakly to peptides shorter than five residues.

One starting sequence for the pools used in this strategy is XX⊗⊗XX, where X represents an equimolar mixture of all 20 amino acids and ⊗ represents defined amino acids. Such a library could be held on 400 pins, or a little more than four blocks of pins. This is a manageable number compared to a library in which three positions are defined, requiring 8000 pins (8^4 blocks). Each of the peptide pools in a XX⊗⊗XX library contains 160,000 members ($20 \times 20 \times 1 \times 1 \times 20 \times 20$). After the initial screen, the dipeptide pool resulting in the most potent antibody binding could be used as the basis for a second-generation library of 400. Identification of the most potent among this set of peptide pools (each containing $20 \times 1 \times 1 \times 1 \times 1 \times 20 = 400$ members) permits two more positions in the peptide of optimum activity to be fixed. A third-generation library of 400 pools would fix the remaining residues. This method is an early representative of the principle now known as iterative deconvolution.

Since molecules discovered by such an approach might have no real relationship to the part of the protein that was recognized by the antibody, but would have the correct size, shape, and charge to mimic the true epitope, they were called mimotopes, meaning mimic of epitope. The mimotope strategy to identify a peptide with high binding affinity to a protein without knowledge of the sequence that it recognizes was tested with an antibody that binds to the known linear epitope DFLEKI. The series of syntheses and screens required to rediscover this sequence is shown on the following page. The number of pins used for peptide synthesis is far fewer than the total diversity of a hexapeptide.

$$XX \otimes \otimes XX \xrightarrow{\text{400 pins}} XXLEXX \Rightarrow \begin{matrix} X \otimes LEXX \\ XXLE \otimes X \end{matrix} \xrightarrow{\text{40 pins}} \begin{matrix} XFLEXX \\ XXLEKX \end{matrix} \Rightarrow \begin{matrix} \otimes FLEXX \\ X \otimes LEKX \end{matrix}$$

$$\begin{matrix} 20 \times 20 \times 1 \times 1 \times 20 \times 20 \\ = 160,000 \end{matrix} \qquad\qquad \begin{matrix} 20 \times 1 \times 1 \times 1 \times 20 \times 20 \\ = 8,000 \end{matrix} \qquad\qquad \begin{matrix} 1 \times 1 \times 1 \times 1 \times 20 \times 20 \\ = 400 \end{matrix}$$

$$\xrightarrow{\text{40 pins}} \begin{matrix} DFLEXX \\ XFLEKX \end{matrix} \Rightarrow \otimes FLEKX \xrightarrow{\text{20 pins}} DFLEKX \Rightarrow DFLEK \otimes \xrightarrow{\text{20 pins}} DFLEKI$$

$$\begin{matrix} 1 \times 1 \times 1 \times 1 \times 1 \times 20 \\ = 20 \end{matrix} \qquad\qquad\qquad \begin{matrix} 1 \times 1 \times 1 \times 1 \times 1 \times 1 \\ = 1 \end{matrix}$$

Peptide synthesis strategy to discover mimotopes

One drawback of the Geysen pin method is that assays are limited to binding processes that can occur while the peptides are covalently linked to the pin. Some assays, such as enzyme inhibition, are not directly amenable to this format. While methods were developed to cleave peptides from pins, other parallel peptide synthesis techniques can be equally attractive if the spatial definition of pins is not used.

Other iterative deconvolution strategies

It is important to recognize that not all peptide coupling reactions proceed at the same rate. Therefore, coupling of equimolar mixtures of the 20 amino acids at the undefined positions in the mimotope strategy would certainly not lead to an equimolar amount of each of the 20 amino acids at that position. Those amino acids that react rapidly would be over-represented, and those that react slowly would be under-represented. If there were a sufficiently large difference in reaction rate, the slow reacting amino acids might not be present at all at those positions where mixtures were used. This could clearly lead to faulty conclusions concerning the best sequence.

Three strategies are available to create equimolar representation of each amino acid when preparing pools. The most obvious is to use a molar amount of the mixed amino acids that is stoichiometric with the amino groups on the support. That is, each of the 20 amino acids is 5% of the mixture, but the total number of carboxyl groups from all of the amino acids in the mixture is equal to the number of amino groups on the support. Provided the reaction is taken to completion, slow-reacting amino acids can "catch up" to fast-reacting amino acids. The drawback of this strategy is that it is difficult to take reactions to completion with only stoichiometric amounts of coupling reagents, which defeats one of the attractive features of SPPS. High concentrations of the reagents in solution can be used in SPPS because the excess can be readily removed by filtration. Because of this high concentration, reactions are kinetically independent of (that is, zero order in) the reagents in solution. Reactions are pseudo-first order in the amino group on the support. They therefore proceed exponentially with time to 100% conversion. This is contrasted with a stoichiometric reaction that is first order in each reagent, which slows as the reaction approaches completion because reagents have been depleted to a low concentration. It is also not very practical to determine the molar quantity of amino groups on the support prior to each coupling reaction.

1. Stoichiometric equimolar reagent mixture

2. Excess isokinetic reagent mixture

3. Split support/couple pure reagents/mix supports

Strategies to obtain equal representation of building blocks in solid-phase synthesis

The second strategy is to adjust the molar amounts of the amino acids in the mixed coupling reagent to compensate for their different reactivity. That is, slow-reacting amino acids are over-represented in the reagents mixture, and fast-reacting amino acids are under-represented in the reagents mixture. Again, considering the kinetics, the rate of coupling of a particular amino acid is the product of its rate constant and its concentration. Even though the coupling to the amino group is zero order in amino acid, the competition between amino acids in reacting with that amino group is still first order. Setting the rate for amino acid 1 equal to the rate for amino acid 2 gives the following relationships.

$$k_1[\text{AA1}] = k_2[\text{AA2}]$$

or

$$k_1/k_2 = [\text{AA2}]/[\text{AA1}]$$

The ratio of amino acids in the mixture should therefore be adjusted by the inverse of the ratio of their coupling rates. Mixtures of reagents of different reactivity that have been created so that reaction rates are nearly identical are called *isokinetic mixtures*. Isokinetic mixtures can be based on direct measurement of reaction rates or simple testing of mixtures of amino acids. Vast experience with SPPS makes the slow- and fast-reacting amino acid residues no mystery. Reasonable judgments can therefore be made of the direction to adjust their concentrations. Evaluation of the ratio of products resulting from the coupling of such a mixture can direct further adjustments in concentrations until an equimolar mixture of coupling products is obtained. Isokinetic mixtures could be determined in a similar way for any other type of reaction that was intended for use in combinatorial library synthesis. This method will be discussed in greater detail in a later section.

The final strategy to obtain equimolar representation of amino acids in mixtures is the split/couple/pool or split/couple/mix method. Split/couple/pool synthesis offers many other virtues that will also be discussed. Important initial concepts of "split synthesis" were contributed by Furka, and its applications were developed simultaneously in several laboratories.

We will first work through an idealized version of split/couple/pool synthesis and then cover the practical aspects. In this case, we wish to prepare a 2-residue combinatorial library from the building block set {X,Y,Z}. That constitutes 3^2 or 9 possible compounds. We need nine support beads to carry the nine compounds, one on each bead. The beads are shown schematically with a dangling bond waiting for attachment to one of the building blocks. Each bead is shown in a different color so that its fate can be traced during this process. The nine beads are split into three pools, and each pool is coupled with one pure building block. Since only one of the reagents is present, excess can be used without concern for relative reaction rates. Each reaction can be taken to completion to maximize the purity of the synthesized product. After this step, each pool contains multiple beads that have only a single building block attached. The three pools of beads are then mixed. A key point is that it is the beads that are mixed, not the compounds. To mix pure compounds is generally anathema to the synthetic chemist, since later separation is often a very difficult process. However, separation of a sort is maintained by each compound's attachment to a different bead, and they can often be used in this form.

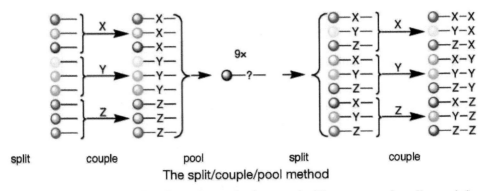

split   couple   pool   split   couple

The split/couple/pool method

This conglomeration is split to give again three pools. The purpose of pooling and then immediately splitting can be readily seen by comparing the composition of the pools before and after this process. Before, each pool contains only one specific building block at the first position. After, each pool contains each of the possible building blocks at this position.

These pools are coupled with one of the three building blocks, creating the sequences within each pool that are shown. All possible 2-residue sequences for this building block set are represented. Note that while 9 different compounds were created in this synthesis, only 6 coupling steps were performed. This is a direct result of the parallelism of combinatorial solid-phase synthesis, in that multiple coupling processes are being performed simultaneously in a single reaction vessel. As was discussed earlier, the number of possible compounds is:

$$N^I, \text{ where}$$

$$N = \text{\# of building blocks and } I = \text{\# of positions.}$$

Generally, the number of steps in this type of synthesis is the product of the length of the molecules being prepared and the number of building blocks:

$$N \times I$$

The use of these equations to calculate the number of compounds and the number of steps in this example is straightforward. At the end of this synthesis, all solid-phase synthesis beads could be combined and the compounds could be removed from them. The result would be a mixture of 9 compounds where each had equal representation.

What features of this example make it idealized? The first idealized feature is that the splitting is assumed to lead to exactly the same number of beads in each pool. If there are not three beads in each pool, their number will be insufficient to carry all three of the possible compounds with that particular building block at that position. Splitting might be performed manually to assure this, and indeed a manual process is involved in the directed sorting techniques that will be discussed later. However, split/couple/mix libraries usually use a stochastic method to split the beads and therefore must use an excess of beads over the number of possible compounds to ensure that there are enough beads in each pool to carry all possible library members. Anywhere from 2×-10× excess of beads is usual.

The second idealized feature is that each of the beads in one pool in the second round of synthesis is assumed to have derived from each of the pools in the first round of synthesis. This is statistically likely, but other scenarios also have a significant probability. Poisson statistics govern the probability of a bead from each previous pool being represented in a current pool. They are a function of the ratio of the number of beads to the library size (the Poisson mean). Again, the use of an excess of beads over the number of compounds to be synthesized is the best way to ensure that there is at least one bead carrying each compound. A 5× excess results in 98% of all possible library members being represented in a pool. A 10× excess gives a negligible probability of any library member being omitted. Given this advantage, one might wonder why anything other than a large excess would be used. Reagent consumption and the overhead in tracking and analyzing product beads are prime considerations.

for Poisson mean = 1
(1-fold coverage)

$P_0 = 0.37$ (37% of compounds omitted)
$P_1 = 0.37$ (37% of compounds present once)
$P_2 = 0.20$ (20% of compounds present twice)
$P_3 = 0.05$ (5% of compounds present thrice)

Poisson statistics in split/couple/mix synthesis

The split/couple/mix method has another important characteristic: even though a collection of compounds is created that contains all possible combinations, each bead within the library bears only one of these combinations. These libraries are therefore also called one bead/one compound libraries. Contrast this situation with the outcome of a synthesis using one of the other methods to obtain equimolar representation of building blocks. If an isokinetic mixture of X, Y, and Z were available, only two synthesis steps would be required to create all possible 2-residue compounds. However, each bead would bear any and all combinations, because different functional group sites on the same bead could react with any of the reagents X, Y, or Z in each step. If the compounds are removed from the bead for testing as a mixture, choice of the isokinetic mixture approach or the split/couple/mix approach has little consequence. However, it has proven very useful to leave compounds attached to the bead during testing, or to otherwise maintain the structural homogeneity of compounds created on a single bead in a split/couple/mix synthesis.

Examples of split/couple/mix peptide libraries

Two different methods for epitope mapping based on split/couple/mix synthesis demonstrate divergent approaches to determining the optimum sequence. The mimotope method prepares peptide pools and tests them while they are attached to the solid phase. One of the split/couple/mix methods prepares pools but tests them in solution, while the other prepares pure peptides but tests them while they are attached to the solid phase.

Houghten used an iterative synthesis and deconvolution approach in which some of the positions in the peptide are synthesized as a mixture of amino acids. However, he performed his synthesis on conventional polymer support beads, so the peptide sequence (or partial sequence) could not be known based on physical location. The resulting peptide pools were released from the beads for testing in solution. Many types of assays beyond epitope scanning that might be desired cannot be performed while the compounds are bound to a solid phase. Release of library compounds into solution makes them available for essentially any type of assay that can be applied to pure compounds, making the library more versatile.

The creation of these peptide mixtures uses a technique developed by Houghten for parallel peptide synthesis, the "tea bag" method. SPPS beads that are kept physically isolated in mesh packets can be treated batch-wise for the common steps of peptide synthesis and individually for coupling of different amino acids. The amino acids Cys and Trp were omitted in this synthesis design because of their chemical sensitivities that make handling peptides containing them more difficult. The first synthesis created 324 pools (18^2):

1. Perform four cycles of synthesis, where each cycle involves preparing 18 mesh bags of resin, coupling each to one pure amino acid, and removing the resin from each bag for mixing and splitting;

2. Split the resin into 18 pools for coupling of the penultimate amino acid, keeping these resins separate;

3. Split the resin from each one of these 18 pools into 18 further pools for addition of the final, N-terminal residue, keeping these resins separate.

The number of compounds in this library is about 34 million. With a library this large, it cannot be proven that all have been made, nor can they be purified (at least using classical techniques, such as chromatography) before testing. Since a pool is a mixture of compounds, purifications that rely on anything other than a property common to all pool members risk fractionating the pooled products and skewing the equimolar mixture of products that such great care was taken to create. Within each final pool the first two amino acid positions in the peptide are fixed and known. The pools created using this method were removed from the support by HF cleavage. Binding to an antibody was determined by competing each pool with a peptide that was already known to bind to the antibody.

$$\text{Ab·DVDPYA} \underset{\text{DVDPYA}}{\overset{\text{DVDPYA}}{\rightleftharpoons}} \text{Ab} \underset{\text{Ac-YPYDVPDYASLRS-NH}_2}{\overset{\text{Ac-YPYDVPDYASLRS-NH}_2}{\rightleftharpoons}} \text{Ab·Ac-YPYDVPDYASLRS-NH}_2$$

Competition binding assay

Using this process, one peptide **DVXXXX** from among the 324 in the first round of synthesis and testing was selected for further elaboration. A second round of synthesis and screening of 18 pools (**DV⊗XXX**: first two positions fixed as DV, third position varied in the 18 pools, remaining positions a mixture of 18 amino acids) permitted an additional position to be fixed. Three further rounds of iterative deconvolution identified the peptide sequence involved in binding to the antibody.

Recall that the convention in describing peptides is that the *N*-terminus is written to the left. Also recall that peptides are synthesized beginning with the amino acid at the *C*-terminus and moving toward the *N*-terminus. These factors combine to make iterative deconvolution work from left to right in the peptide sequence as written. A drawback to this method is that it is a serial process in which peptides must be made 5 times.

⊗⊗XXXX $\xrightarrow[\substack{\text{peptide} \\ \text{mixtures}}]{\text{test 324}}$ DVXXXX ··➤ DV⊗XXX $\xrightarrow[\substack{\text{peptide} \\ \text{mixtures}}]{\text{test 18}}$ DVPXXX ··➤ DVP⊗XX $\xrightarrow[\substack{\text{peptide} \\ \text{mixtures}}]{\text{test 18}}$

DVPDXX ··➤ DVPD⊗X $\xrightarrow[\substack{\text{peptide} \\ \text{mixtures}}]{\text{test 18}}$ DVPDYX ··➤ DVPDY⊗ $\xrightarrow[\substack{\text{peptide} \\ \text{mixtures}}]{\text{test 18}}$ DVPDYA

Iterative synthesis and deconvolution of a peptide library

An alternative way to use the split/couple/mix method performs all synthesis at the beginning of the experiment, creating the complete set of peptides of a given length, one sequence per bead. The library can be a continuing resource for biological screening against any number of targets. Like the mimotope method, it assays the peptides while they are attached to the solid phase, but unlike the mimotope method, it does not use mixed pools.

The power of the split/couple/mix method for creating a peptide library can be appreciated by a thought experiment. Consider a pentapeptide library. The number of sequences is

$$20^5 = 3.2 \times 10^6$$

The synthesis must therefore begin with at least 3.2 million beads, which are split into 20 pools. Splitting can be accomplished using a DMF/CH$_2$Cl$_2$ mixture that has the same density as the beads. Beads are suspended in this *isopycnic* solution, and can be partitioned by a volumetric pipette. Each pool is coupled with a single activated amino acid. The pools are combined, the *N*-terminal protecting group is removed from all of the beads, and they are again split into 20 pools. The process is repeated four times to create a library of pentapeptides in which each bead bears but one sequence.

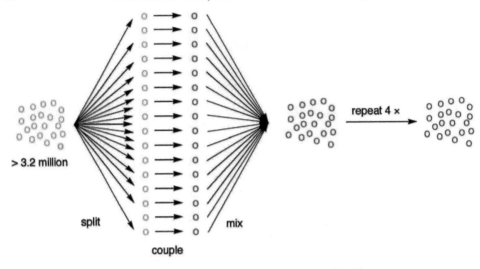

Split/couple/mix synthesis of a pentapeptide library

The chemical efficiency of this synthesis is amazing. It generates 3.2 million compounds in 100 chemical coupling processes, or 32,000 compounds per step. For this sort of oligomeric molecule, the library size increases with the power of the length, but the chemistry required to make it is only the product of the length and the number of building blocks. Recall that the diameter of synthesis beads is ~100 μm. This enables a large collection of compounds to be contained in a small volume. An estimate of library volume can be calculated.

5 (× 20) chemical couplings
(a linear function of the building blocks)

give

3.2×10^6 compounds
(an exponential function of the building blocks)

3.2×10^6 compounds × 10-fold library coverage × $\pi(5 \times 10^{-3} \text{ cm})^3 = 4 \text{ cm}^3$

Power of combinatorial synthesis

As a model system to demonstrate an on-bead assay method, Lam used epitope mapping of an antibody that had been raised against the bioactive peptide β-endorphin, an opiate peptide from the brain with the sequence YGGFL at its N-terminus. The 3E7 antibody was already known to recognize the five amino acids at the amino terminus of the β-endorphin sequence. For synthetic convenience, 19 amino acids (omitting Cys) were used to prepare this library, making it a bit smaller (19^5) than the full pentapeptide library. Testing was performed using a novel binding assay.

The ability of an antibody to bind to a peptide sequence is evidently not impaired by its being attached to a solid phase, as was shown in the peptide-on-pins work. Antibody binding assays are called ELISAs, for enzyme-linked immunosorbent assay. In bead experiments, they involve coloring those beads that bind antibody. An antibody (such as a mouse monoclonal antibody) solution is added to a solid support that bears a peptide, and through antigen-antibody recognition, becomes bound to the support. A protein conjugate is obtained between an antibody that binds to a constant region of the first antibody (e.g., a goat anti-mouse antibody) and an enzyme that can produce a colored product (such as indigo) in the presence of a specific substrate. This enzyme-linked secondary antibody is added and is thereby also bound. Finally, the specific substrate is added and the colored product indigo is produced by the enzyme. Because it is insoluble, indigo precipitates on the surface of the bead, coloring it. If we apply an ELISA to the whole library of beads, some will bind primary antibody and some (most) will not. If they do bind primary antibody, they will bind secondary antibody and develop a color when substrate is added. These beads can be physically selected under a microscope. Beads are extensively washed to remove all of the different types of proteins used in the assay, then sequenced.

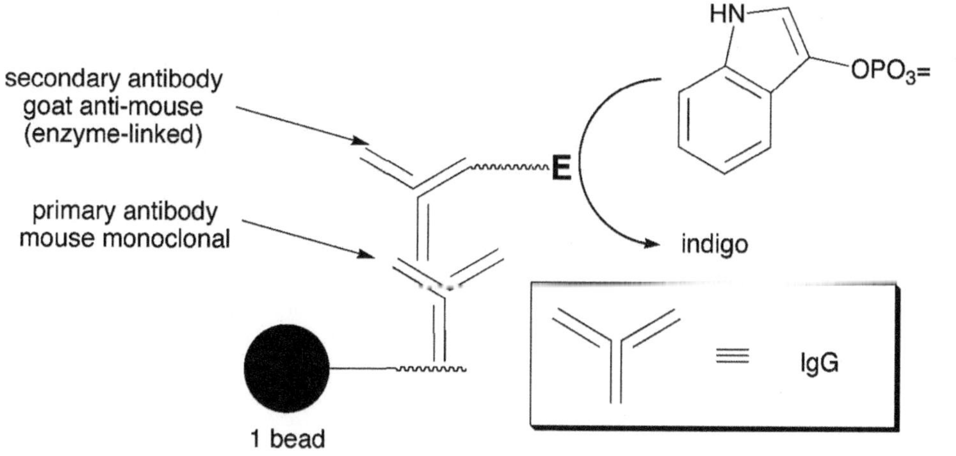

On-bead ELISA assay

With beads that bind antibody identified, the sequences of the peptides on those beads must be determined. It is here that the one bead/one compound nature of these libraries is crucial, since all peptides on a bead have the same sequence (barring synthesis failures). The identification of the sequence that binds antibody can be performed by microsequencing using the Edman degradation method. Treatment with aryl isothiocyanates generates phenylthiohydantoin (PTH) derivatives of the amino acids, beginning at the N-terminus, as shown on the following page. Commercial instruments can automatically identify PTH derivatives by HPLC with picomolar sensitivity. Recall from the earlier discussion that the number of functional groups on a 90 μm bead is about 100 pmol, well above this sensitivity.

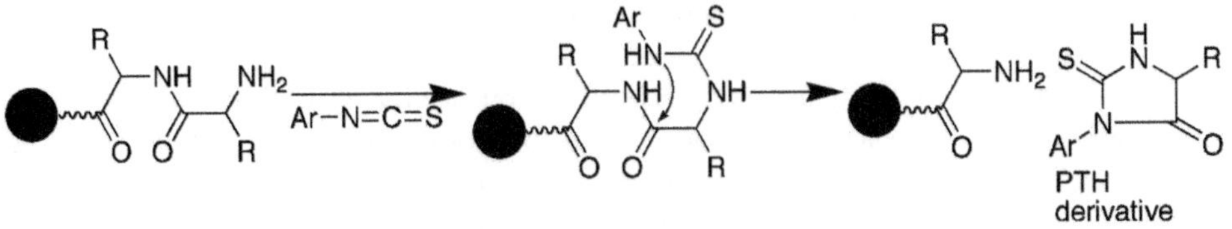

Edman degradation peptide sequencing

While each bead bears only one peptide, the analysis of active beads in an epitope scanning experiment is not expected to give only one sequence. Perhaps antibody binding does not involve some positions in the peptide, or the antibody cannot easily distinguish one amino acid from another (such as Ile/Leu/Met). A *family* of sequences is the expected outcome from this experiment. Using the foregoing methods, Lam was able to verify that the YGGFL peptide sequence is recognized by antibody 3E7, as well as discover several novel sequences that are recognized as well.

Even the venerable Edman degradation has been supplanted by more modern techniques for determining peptide sequence. Mass spectrometry is frequently used to determine the overall molecular weight of a peptide, from which the amino acid composition can be determined. Mass spectra with collision-induced dissociation (CID) can generate fragment ions of peptide subsequences. The amino acid composition of these subsequences can also be determined, and the overall peptide sequence can be assembled logically.

A powerful technology for analysis of biomolecules that works well with peptides is MALDI-TOF mass spectrometry. MALDI stands for matrix-assisted laser desorption ionization, and TOF stands for time-of-flight. A powerful laser impinges on a steel plate on which is spotted the sample and the matrix. The matrix is simply a molecule that absorbs laser energy and transfers it to the analyte, putting the molecule into the gas phase and ionizing it, generally by protonation to give an MH^+ ion. The ions are accelerated by a fixed potential. As is usual in mass spectrometry, heavy ions have a slower velocity and lighter ions have a faster velocity. The light ions arrive at the detector earlier, the heavy ions later. Hence, the mass separation is based on their time of flight.

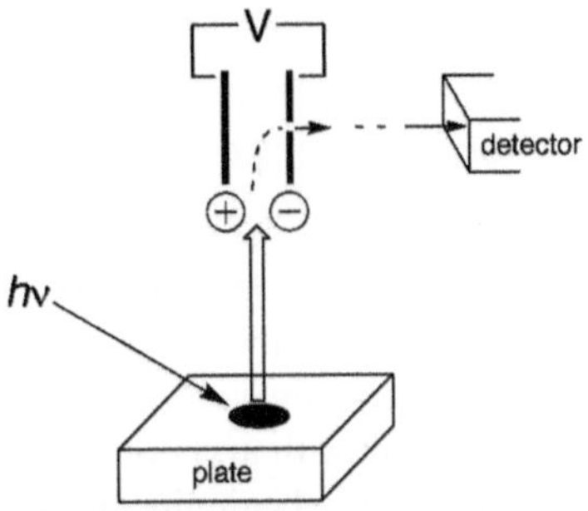

MALDI-TOF mass spectrometry

Conventional MALDI-TOF MS might be used to sequence the peptides identified in an on-bead assay, but advantage can be taken of the fact that the peptides in a combinatorial library are synthesized. Rather than relying on CID to provide fragments that lead to the sequence, the sequencing-by-synthesis method (also called ladder sequencing) can be applied to identify the peptide directly.

The general concept of sequencing-by-synthesis involves the creation of a full-length version of the molecule whose sequence is to be determined as well as a nested set of truncation molecules. A comparison of the sizes of the full-length and successive truncation molecules identifies the residue at each position. For example, the dideoxy chain-termination sequencing method for DNA sequencing is a representative of the sequencing-by-synthesis method. Sequencing-by-synthesis is particularly applicable to DNA because DNA can be replicated by polymerase enzymes and dideoxynucleotides are accepted in place of deoxynucleotides during replication, thereby terminating the chain. A low concentration of one dideoxynucleotide terminator is used with the natural deoxynucleotides to create molecule sets that are truncated only at the positions corresponding to that base. The size of the truncated DNA molecules is determined by polyacrylamide gel electrophoresis (PAGE), which cannot distinguish the individual DNA bases. In its simplest manifestation, the Sanger method therefore requires one PAGE for each of the four nucleotides.

A very similar concept can be applied to peptide sequence determination, with the exception that enzymes are not available that can replicate peptide sequence. As an example, consider the peptide sequence YGGFLQ recognized by the 3E7 antibody. If the full-length peptide were analyzed by mass spectrometry and compared to the n-1 truncation peptide, they would differ by 163 Da, the mass of the terminal tyrosine residue (minus the water released from amide bond formation). This identifies tyrosine as the N-terminal residue. Likewise, the amino acid following it in the sequence can be identified by the mass difference with the n-2 peptide.

$$
\begin{array}{l}
\text{Y—G—G—F—L—Q- -} \\
\text{G—G—F—L—Q- -} \\
\text{G—F—L—Q- -} \\
\text{F—L—Q- -} \\
\text{L—Q- -} \\
\text{Q- -}
\end{array}
\left.\begin{array}{l}
\\ \Delta = 163\ \text{Da} \\ \Delta = 57\ \text{Da} \\ \Delta = 57\ \text{Da} \\ \Delta = 147\ \text{Da} \\ \Delta = 113\ \text{Da}
\end{array}\right.
$$

Masses of nested truncations of a peptide uniquely identify its sequence

The source of the set of truncation molecules is key to the method. They can be created during the peptide synthesis by inclusion of a small amount of an N-acetylated amino acid in the coupling with each N-Fmoc-protected amino acid. While the Fmoc is deprotected in the subsequent step, the acetyl group is stable to all conditions used during the remainder of the synthesis. Thus, 5% of the growing peptide chains are truncated at this residue. Binding assays can be reliably performed on beads bearing low amounts of truncation sequences, with the binding activity being ascribed to the dominant, full-length peptide and the truncation peptides providing an immediate method of structural determination. Once beads with desired sequences are identified, the peptides must be removed from the support for analysis by MALDI-TOF MS. Conveniently, the bead itself can be used to carry the peptide after cleavage. A linker that is sensitive to TFA is used to begin the peptide synthesis. Dry beads exposed to TFA vapor will undergo cleavage of the attachment to the peptide, but it will remain adsorbed to the bead surface. The bead can be directly subjected to mass spectrometry, or the peptide can be first eluted. The mass spectrum of this sample will give a parent ion and mass difference signals corresponding to the truncations of individual amino acids. A potential flaw of this method is that it cannot distinguish the isobaric amino acids leucine and isoleucine. However, this problem can be readily addressed by using a cap with a slightly different mass on one of the two.

Partial termination of the Gly-Gly coupling in peptide synthesis facilitates sequencing

When a library has been made by the split/couple/mix method with an excess of beads over the number of compounds, the frequency at which a particular sequence is detected should be governed by the same Poisson statistics as the likelihood of a given compound appearing in the library. Thus, for a library prepared with a 10× coverage of the calculated diversity, hits will appear on average 10 times. This means that sequencing must be performed on 10 times as many beads as result in valid hit compounds. If a hit is not detected with a statistically appropriate frequency, it cannot be valid. Unlike combinatorial synthesis, sequence analysis is not generally a parallel process, so it can add significant time to a library experiment. This is an additional reason to limit the excess coverage in split/couple/mix libraries, keeping in mind the risk that some compounds imputed from calculating the diversity may not actually be present.

Positional scanning

While the foregoing methods are powerful, a technique is still needed that can be used with any type of biological assay (not limited to solid-phase binding assays) and avoids the circular process of synthesis and screening in iterative deconvolution. One such method is positional scanning. It is similar to iterative synthesis and deconvolution in that it uses pools with defined positions and mixed positions. However, all of the pools are synthesized at one time, and their screening is intended provide a direct readout of preferred peptide sequence(s). A hexapeptide positional scanning library has the following form:

$$⊗XXXXX$$
$$X⊗XXXX$$
$$XX⊗XXX$$
$$XXX⊗XX$$
$$XXXX⊗X$$
$$XXXXX⊗$$

X again represents an equimolar mixture of all 20 amino acids, and ⊗ represents pools with a known amino acid at that position. There are 20 pools for each position, or 120 peptide mixtures overall. Testing of the 20 first-position pools should identify the superior first position amino acid, since that is the only difference among the different pools. Likewise, the superior amino acids at each of the other positions can be identified, and these can be assembled into the presumptive best sequence overall.

Positional scanning libraries are prepared using the isokinetic mixture method and thus are not one bead/one compound libraries. Since positional scanning libraries are intended to be removed from the support for testing, the fact that beads bear multiple sequences does not pose a problem. To appreciate why isokinetic mixtures are used, consider the creation of the sixth-position positional scanning library. That is, the amino acid at the (right hand) C-terminus that is the first residue coupled in the synthesis is defined. Twenty packets of peptide synthesis resin can be treated with one of the amino acids. Those packets can be put into the same vessel for deprotection of the Boc groups, but they must be kept separate for the remainder of the synthesis (20 pools × 5 coupling steps) in order to maintain the identity of the sixth-position amino acid. While theoretically possible, performing split/couple/mix on each of these initial sub-libraries would mean 100 additional coupling steps on each sixth-position sub-library, or 2000 couplings, which is impractical. Therefore, the isokinetic mixture of amino acids is used for the addition of subsequent residues.

Generally, an attraction of the isokinetic mixture method is that it scales only with the length of the molecule to be made and therefore requires many fewer coupling steps than split/couple/mix, which scales with the product of the length and the number of building blocks. A hexapeptide synthesis requires only six steps using an isokinetic mixture, and is independent of the number of building blocks. A hexapeptide synthesis requires 120 steps using split/couple/mix, and if the 20 (R)-isomers of the genetically coded amino acids were added to the library, the number of steps would increase to 240. Of course, the determination of the isokinetic mixture for a synthesis can be time consuming, and must be repeated if any change is made to the building block set. This would not be necessary if the change involved only the addition of enantiomers, of course, since they should have nearly identical coupling rates. Another difficulty with isokinetic mixtures is that the rate of coupling is dependent not only on the amino acid that is being coupled, the rate is dependent on the amino acid to which it is being coupled. That is, coupling Val to Val may be slower than coupling Val to Ala, and different amino acids may be affected differently by the amino terminal residue. Because the amino acid to which a residue is being coupled is not usually known in a library synthesis (in this example, it is a mixture of all 20), the creation of isokinetic mixtures is an imperfect science.

A potential problem with positional scanning is the assumption that the most active pool corresponds to the single peptide with the most potent amino acid at its cognate position. This applies as much to the Geysen mimotope method as to Houghten's method. Assay results for pools should be proportional to the product of the concentration of each compound in the pool and its activity, summed over the pool. Ideally, the concentration of each compound in the pool will be the same if the synthesis has gone as planned; this is one reason that using one of the three synthesis techniques to obtain equimolar mixtures is important. It might be possible that a potent pool is the result of a large number of weakly potent compounds rather than a single very potent compound. Alternatively, a potent compound could be present in a pool with many other weakly potent "diluents" that would prevent that pool and therefore that compound from being identified as active.

Using the β-endorphin 3E7 antibody discussed earlier, Houghten demonstrated positional scanning and dealt with its possible flaw by taking the fixed amino acid from not only the best pool but up to four pools for use in finding the optimum peptide sequence. In the case at hand, this gave a building block set for a second round of combinatorial synthesis with the listed residues at the indicated positions for the synthesis of individual peptides, comprising 24 combinations. Those peptides were analyzed for binding to 3E7 and YGGFMY was found to be superior, consistent with previous work with single peptides and peptide libraries.

1	2	3	4	5	6
Y	G	G	F	F	F
		F		Y	Y
				M	R
					L

The results of a pooled screening experiment can be analyzed statistically to determine the likelihood of finding a valid hit. Screening of any pool of compounds is limited by the precision and sensitivity of the assay, the potency of the active compounds, and the number of compounds in each pool. Assay values for a pool are the average of those for each of the compounds in the pool. The potency of an active compound A_{act} must be greater than the pool average by an amount greater than the error in the assay to reliably identify an active pool. Based on this idea, eq 1 can be derived from statistics (the comparison of experimental means, Student's t-test). The number of compounds in each pool is n and the assay standard deviation is σ. In the limit of large n, this expression reduces to eq 2.

$$A_{act} - A_{ave} > t\,\sigma \sqrt{\frac{(2n-1)(n-1)}{n}} \quad (1)$$

$$A_{act} - A_{ave} > \sigma\, 2\sqrt{2}\,\sqrt{n} \quad (2)$$

The second equation reflects properties expected for the screening of pools. Greater potency (A_{act} large) will obviously make actives easier to detect. A significant population of mediocre compounds (A_{ave} large) will have the opposite effect, raising the background and making it more difficult to identify actives. Precise assays (σ small) will make actives easier to find. Lone active compounds are more likely to be missed with larger pools (n → ∞) because they are diluted by inactive components, but this effect rises only with the square root of the pool size. Larger pool sizes may therefore be helpful, since they increase the probability that at least one active is present in a pool. While sensitivity for an active compound is greater at smaller pool sizes, sensitivity decreases more slowly than does pool size. It is impossible to quantify the relationship between pool size and the likelihood of finding an active *a priori*; it will be unique to each possible library chemistry and assay.

Epilogue

Combinatorial chemistry often involves at least some components of making and testing mixtures as opposed to single compounds, non-purified compounds as opposed to purified compounds, and molecules bound to solid phases as opposed to those in solution. Ideally, in finding a new molecule with a desired activity, compounds of known structure and purity would be tested in solution–based assays. However, it is impossible to reach the large numbers of possibilities available (e.g., 34 million) in this way because conventional synthesis and testing usually cannot be done in parallel.

The classical chemical paradigm is:

synthesize → purify → characterize/identify → test,

but with huge numbers of compounds, great resources would be expended purifying and characterizing compounds that are not active in testing. If the synthesis is efficient enough that purification is unnecessary and the identity of each compound is known before testing, or can be easily determined afterward, the paradigm can be inverted to:

synthesize → test → identify.

This has offended many chemical traditionalists who are accustomed to having full analytical data on all new compounds. Resources do not permit these data to be obtained on huge libraries, however. Also, in the examples given, peptide synthesis was used, which represents some of the most reliable and high-yielding synthetic chemistry known. It is reasonable to presume that the vast majority if not all of the compounds intended were in fact prepared, subject to the statistical limitations mentioned.

A useful parallel can be made between combinatorial library testing and natural products screening. It is common to use a fermentation extract, Chinese herb, or Amazon plant, which may contain many thousands of compounds, in screening and bioactivity-guided purification to discover novel agents. No one is concerned about the possible false negatives in this research, nor the fact that impure compounds are tested. Combinatorial chemistry offers the advantage over natural product extracts that the structure of the actives that are found is known, or very readily determined. On the other side of the issue, subsequent research has sometimes found difficulties in tracking down "hits" within combinatorial compound mixtures.

The methods for creation and testing of peptide libraries summarized here, and other methods to be covered in later sections, have established important new paradigms for discovering biologically active compounds. They are also valuable because of the great significance of peptides in biology. While the examples discussed do not reveal the application of these methods to the discovery of novel and interesting peptide sequences, such discoveries have in fact been made. Naturally, modifications and improvements to these methods were required. For example, while antibody binding to polystyrene-bound peptides is successful, other proteins may not recognize their peptide ligands when they are presented in this format. Consequently, novel hydrophilic supports such as the TentaGel® resins were required to permit peptide libraries to be used for drug discovery. However, peptides have drawbacks in pharmaceutical development, particularly regarding their low oral activity, their sensitivity to the proteases that seem to be pervasive, and their rapid excretion. Some of these issues can be addressed by synthetic peptide libraries. For example, because proteases have evolved to recognize peptides composed of (S)-amino acids, (R)-amino acid-containing peptides are resistant to proteases. To a lesser extent, proteases may be unable to hydrolyze cyclic peptides or those containing unnatural amino acids. While there are clearly some important peptide drugs, such as cyclosporin, an anti-rejection agent for organ transplants, peptides are also not the dominant structural class within pharmaceuticals. Principles and methods of combinatorial chemistry drawn from the peptide library area have therefore been applied outside it. One of the prime areas of progress in solid-phase synthesis and combinatorial methods has been in small, drug-like organic compounds.

Additional reading

"Solid Phase Peptide Synthesis," Fields, G. B., Ed., Meth. Enzymol. 289, Academic Press; San Diego (1997).

"Peptide Chemistry. A Practical Textbook," Bodanszky, M. Springer; New York (1993).

Hoffmann, C.; Blechschmidt, D.; Krüger, R.; Karas, M.; Griesinger, C. Mass spectrometric sequencing of individual peptides from combinatorial libraries via specific generation of chain-terminated sequences. *J. Comb. Chem.* **2002**, *4*, 79-86.

Lam, K. S.; Salmon, S. E.; Hersh, E. M.; Hruby, V. J.; Kazmierski, W. M.; Knapp, R. J. A new type of synthetic peptide library for identifying ligand-binding activity. *Nature* **1991**, *354*, 82-4.

Houghten, R. A.; Pinilla, C.; Blondelle, S. E.; Appel, J. R.; Dooley, C. T.; Cuervo, J. H. Generation and use of synthetic peptide combinatorial libraries for basic research and drug discovery. *Nature* **1991**, *354*, 84-6.

Problems

1. List the sequences of the peptide products of the following mix-split synthesis.

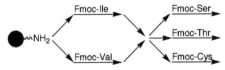

2. A peptide library prepared by the terminated split/pool synthesis method was used to find beads that bind to the protein streptavidin in competition with biotin. The MALDI mass spectrum of one such bead shows a parent ion M^+ and fragment ions M^+-137, M^+-234, and M^+-362. What is its sequence?

3. Calculate the diversity in one pool of the initial peptide library used in the Houghten iterative deconvolution method.

4. A solid-phase peptide synthesis proceeds with a step-wise coupling efficiency of 98%. If a decapeptide is synthesized, what is the maximum percent purity that may be exhibited by the peptide (as analyzed by an appropriate method, like HPLC) immediately upon removal from the support?

5. Describe how the composition of a peptide library composed from (R) and (S) versions of the 20 genetically coded amino acids would differ when prepared on solid-phase synthesis

supports with the isokinetic mixture method and the split/couple/mix method.

6. Write a reaction scheme for the loading of Fmoc-alanine onto Rink amide resin, removal of the Fmoc group, acetylation, and acid-promoted release.

7. Binding of the HIV tat protein to the TAR (transactivating response region) RNA element activates transcription of the HIV viral genome. A peptide that binds to TAR in competition with tat was sought. A library was prepared on 1215 Merrifield resin beads by five rounds of split/couple/mix synthesis with the amino acids R, K, and N, with partial termination of synthesis by N-acetyl versions of each amino acid. Calculate the number of sequences in the library.

The library was screened against a pre-labeled TAR molecule, and beads binding TAR as evidenced by their coloration were selected. Peptide MALDI-TOF ladder sequencing was applied to the binding beads. The following was the outcome of MS analysis of the 16 colored beads, with the molecular and fragment ions detected in the analysis for each. Which are valid hits? What is the consensus sequence for binding of tat to TAR?

3×	M^+, M^+-114, M^+-270, M^+-426, M^+-540
1×	M^+, M^+-114, M^+-228, M^+-384, M^+-512
7×	M^+, M^+-114, M^+-270, M^+-426, M^+-554
5×	M^+, M^+-128, M^+-284, M^+-440, M^+-554

8. Highly cationic peptides have interesting antibacterial properties. We would like to generate and screen by iterative deconvolution a combinatorial peptide library composed from phenylalanine, valine, and lysine, each in both (R) and (S) forms. In model reactions with the BOP coupling agent, the observed coupling rate constants with (aminomethyl)polystyrene are as given below:

Boc-Val	2×10^3 s^{-1}
Boc-Lys	3×10^3 s^{-1}
Boc-Phe	5×10^3 s^{-1}

What mole percentage composition of Boc-protected amino acids should be used for the preparation of this library by the isokinetic mixture method?

9. A peptide library was made using the split/couple/mix method with partial termination of the synthesis. The library was screened for binding to chlorophyll, and green beads were selected. Individual beads were submitted to MALDI-TOF MS analysis. The observed fragments were as follows (parent ion M^+): fragment ions M^+-71, M^+-199, M^+-330, and M^+-477. What is the sequence of the binding peptide?

Supports, Linkers, and Reagents for Peptide Synthesis

Beaded solid-phase synthesis supports are widely available as articles of commerce, and those that cannot be purchased are readily prepared from other supports that can be purchased by using fairly straightforward chemistry. Attachment to the support of starting materials for a synthesis often mimics a protecting group that is commonly used in conventional organic synthesis. This enables the support to play dual roles of carrier for the compounds and one of the protecting groups during the synthesis. Opportunities for new support development can also be discovered in the literature on protecting groups.

Polystyrenes

Polystyrene is commonly used as a support in peptide synthesis, and it has been adapted to many other types of compounds. Typically, it contains 1-2% divinylbenzene cross-linking, and is supplied as beads with a high surface area and interior volume. Polystyrene can be swollen by a number of solvents, including toluene, THF, CH_2Cl_2, and DMF, resulting in a gel phase. Generally, swelling is necessary to provide reagents with access to the interior of the bead. As an industrial product, polystyrene is available in a wide range of grades, many of which contain impurities that at the least will leach out into reaction products, if not cause synthesis failures altogether. It is crucial to know one's polystyrene supplier and to know that the supplier maintains strict quality control. Supports must also be washed before use.

A key feature of any support is its loading, expressed in terms of the number of millimoles of functional groups per gram of resin. Values range from 0.3 mmol/g up to ca. 2 mmol/g, and are commonly about 1 mmol/g, corresponding to 250 mg of a compound of molecular weight 250 Da on 1 g of resin. The loading defines the average distance between functional groups on the support, where lower loading corresponds to increased distance between functionalization sites, and higher loading corresponds to decreased distance between sites. It had been believed for some time that with sparse functionalization of polystyrene and the rigidity of the cross-linked polymer, chemical interaction between different sites of functionalization would not be possible. In favorable circumstances, this feature could be exploited, for example, to selectively derivatize one functional group in a bifunctional molecule. Clearly, this "site isolation" must be a function of the loading, and it has subsequently been shown that with higher loading resins, reactions between sites on the same resin bead are possible. The loading thus has an influence on inter-site reactions on the support that is analogous to the influence of concentration on intermolecular reactions in solution.

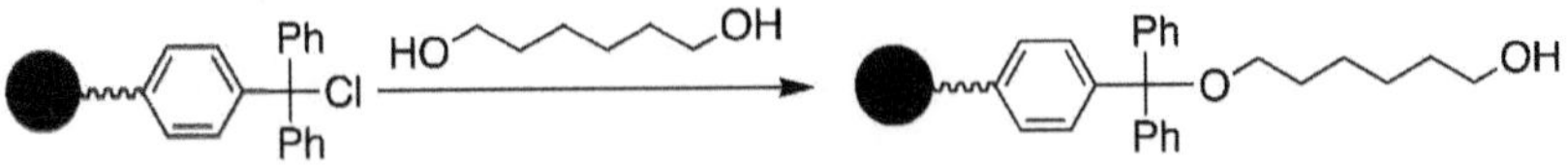

Site isolation on solid phases can permit selective protection

One method to probe the inter-site interactions in resins uses the fluorescence of the pyrene excimer. Pyrene can form an excited state dimer (excimer) that has a longer λ fluorescence emission than the monomer. Benzylamine resin of various loadings can be acylated with 4-pyrene butyric acid. At 1 mmol/g, only excimer emission is seen, while at 0.5 mmol/g, the majority of the fluorescence comes from the monomer.

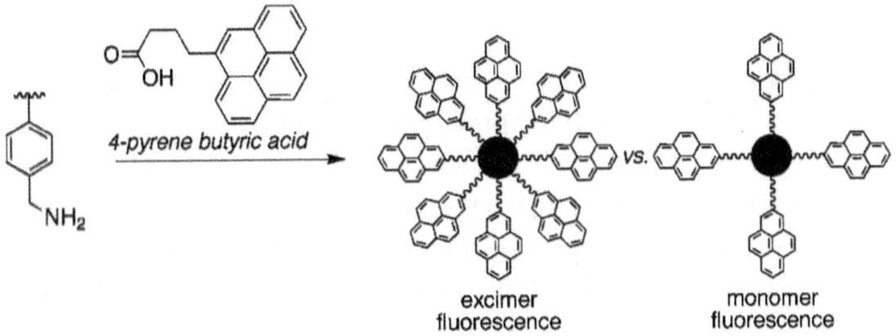

4-pyrene butyric acid

excimer fluorescence   vs.   monomer fluorescence

The simplest functionalization of polystyrene is with a chloromethyl group, originally performed by treatment with HCl and formaldehyde, creating Merrifield resin. It is also possible to copolymerize (chloromethyl)styrene with styrene and divinylbenzene to introduce this functionalization, but this approach suffers from the drawback that commercial chloromethylstyrene is a difficult-to-separate mixture of *m*- and *p*- isomers. Loading of carboxylate salts onto Merrifield resin by displacement of the chloride occurs fairly readily. The resulting product is effectively a benzyl ester, which can be cleaved under fairly strong acid conditions, such as liquid HF.

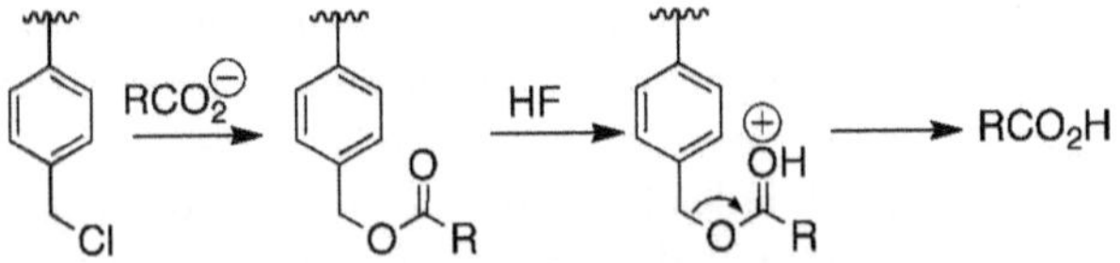

Merrifield resin - loading and removal of an acid

Acidic conditions are often utilized to remove compounds from solid-phase synthesis supports. This strategy was initially developed for peptide synthesis because acidic conditions do not cause racemization of the α stereogenic center of amino acids in peptides. It also offers the virtue that the reactivity of protecting groups for the synthesis cycles, protecting groups for the side chains, and the attachment to the support (a sort of protecting group) can be tuned for removal under different acidic conditions. Given that these removal steps are promoted by acid, the removal reaction is generally Sn1-like, and this tuning of reactivity is most often based on the stability of the resulting carbocation. Thus, addition of electron-donating groups accelerates acid-catalyzed removal. Another virtue of acidic deprotection and removal reactions is that volatile acids (such as formic, acetic, and especially trifluoroacetic acids) are readily available. Volatility permits the reagent used in the deprotection to be simply evaporated rather than requiring a work-up step, with its attendant material losses.

A simple modification to Merrifield resin makes it easier to remove products with acidic cleavage. Wang resin is obtained by displacement with *p*-hydroxybenzyl alcohol, creating a resin equivalent of *p*-methoxybenzyl alcohol. Acids are loaded onto Wang resin as their active acylating agents, and can be cleaved from the support with TFA.

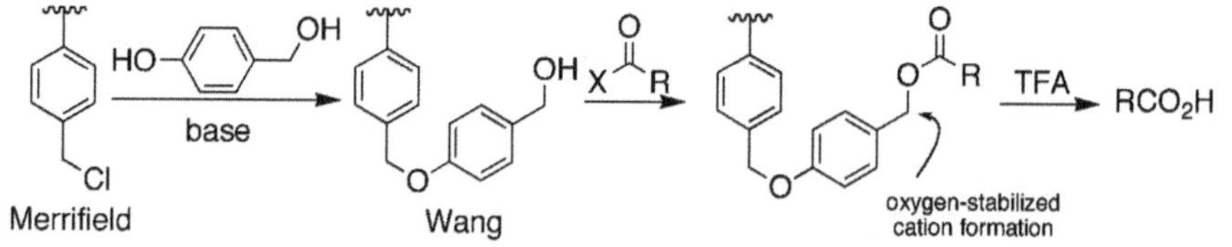

Wang resin - loading and removal of an acid

Beginning with a rudimentary resin, additional molecules, often called "handles," can be added to impart different chemical properties (and cleavage conditions) to the support-bound molecules. For example, very mild acidic conditions for release of an acid from the support are obtained with the Rink ester linker. Rink ester supports are prepared by acylation of (aminomethyl)polystyrene resin with an aryloxyacetic acid derivative. A corresponding Rink amine is available that produces the amide. Of course, mild conditions for cleavage of the product from the support mean that the strength of acidic reagents that can be tolerated while the compound is intended to remain on the solid phase is more limited.

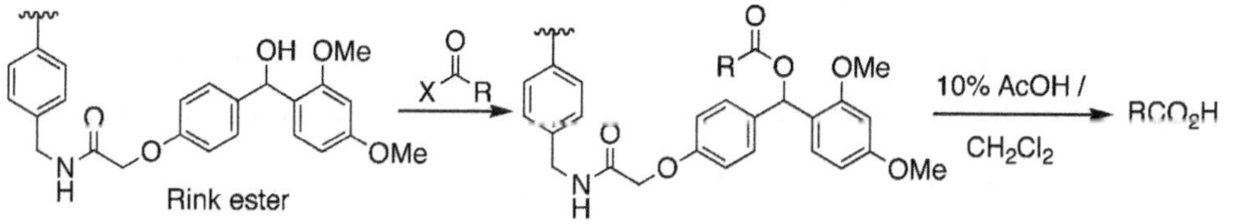

Rink resin - loading and removal of an acid

The Kaiser oxime resin provides an example of a base-promoted removal process. Here, the acid to be loaded onto the support is esterified with the hydroxyl group of a benzophenone oxime. This linkage tolerates Boc-based peptide synthesis, but is reactive as a leaving group in an acyl substitution reaction due to extensive delocalization of the oxime anion. This support

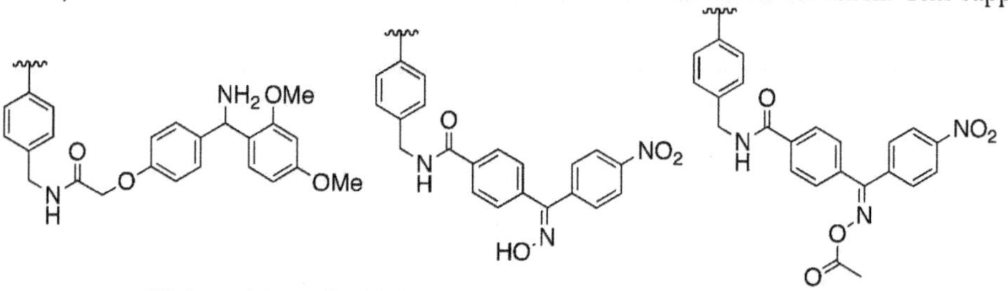

Rink amide resin, Kaiser oxime resin, and a Kaiser resin ester

was initially developed to permit peptide segments prepared by solid-phase synthesis to be removed from the support while maintaining their side-chain protection. One nucleophile used for this purpose is N-hydroxysuccinimide (NHS), which forms the "active ester," a derivative that is reactive for peptide bond formation. These protected/activated peptide segments can then be purified. Kaiser resin provides access to pure peptides protected at their side-chains and activated at their C-terminus. Combination of such segments permits the assembly in pure form of much larger polypeptides than could be prepared by a linear synthesis.

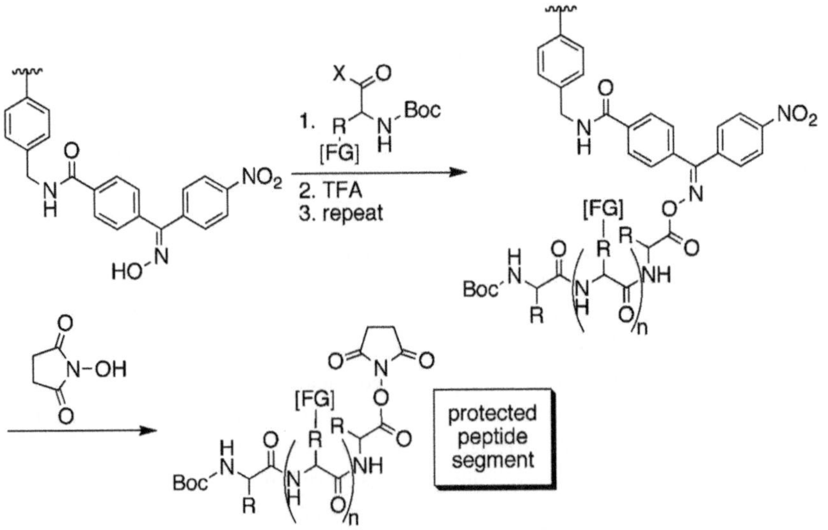

Synthesis of a protected peptide segment on Kaiser oxime resin

In general solid-phase synthesis techniques, Kaiser resin oxime esters are reactive enough that acyl substitution can be performed with only one molar equivalent of nucleophile, such as an amine. Any ester (or ester-based linkage to a support) can in principle be treated with an amine to form an amide in an aminolysis reaction. This reaction is generally favorable due to the greater thermodynamic stability of amides. For ester-linked supports, formation of the amide removes the compound from the support. However, practically, the rate of this reaction may not be sufficiently favorable to permit it to be performed with only one equivalent of amine. This factor is important when non-volatile amines are used so that difficulty with removing excess amine is not encountered.

Photochemistry is another common strategy for release of compounds at the end of a synthesis. It offers the virtues that it is quite orthogonal to most other types of chemistry. The chance of accidentally releasing the compound from the support by exposure to light is low, provided that care is taken in the selection of the protecting group and the handling of the resin during synthesis. No reagents remain from the deprotection reaction that must be removed from the product. The Holmes photolinker is based on some well-known nitrobenzyl photochemistry. Ester derivatives of this photolinker can be cleaved with 365 nm light. This wavelength is commonly used to visualize thin-layer chromatography plates. The related amine can be used to give amides after photochemical release.

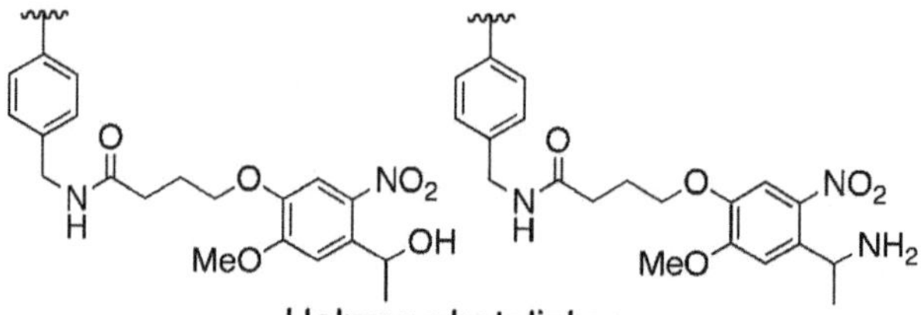

Holmes photolinkers

PEG-grafted supports

Polystyrene supports have significant drawbacks in any synthesis or screening that involves water or other non-swelling solvents. It has been found that while antibodies will bind to peptides linked to polystyrene, other proteins will not. A more hydrophilic replacement for polystyrene was therefore needed in order to broaden the types of protein targets that can be screened for binding with synthetic peptide combinatorial libraries. Such supports were already known, and many are commercially available. Polyethyleneglycol-grafted polystyrenes, such as TentaGel® and ArgoGel,® can be prepared from Merrifield resin in one Sn2 displacement step, or a few steps. The resulting supports swell in most solvents, including water. Another virtue of these supports is the extremely long linker arm of the polyethyleneglycol chain that extends far away from the polystyrene bead and might offer a more "solution-like" environment. The X groups at the ends of the polyethyleneglycol chains can be simple amines or alcohols, or they can be one of the handles described above that impart specific chemical reactivity for cleavage and create specific products upon cleavage. One drawback of these grafted supports is that their loading can be relatively low, about 0.25 mmol/g.

The sizes, loadings, and number of beads per gram for several TentaGel supports are summarized in Table 2. Beads larger that about 200 μm are sometimes called "big beads," but their size is still relatively small - comparable to a salt grain.

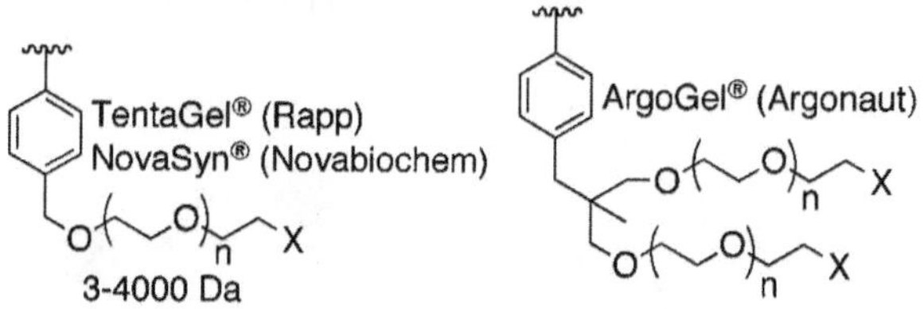

PEG-grafted polystyrene supports

Table 2. Physicochemical Properties of TentaGel Beads

Diameter	Count	Loading
90 μm	2,900,000/g	100 pmol
130 μm	890,000/g	300 pmol
250 μm	110,000/g	2300 pmol

Coupling strategies

While peptide bonds are not the most challenging bonds to form in organic synthesis, the methods used to form them have been important to the advancement of combinatorial chemistry precisely because they are so reliable. *N*-Protected amino acids are traditionally activated *in situ* for peptide coupling. In so doing, another feature of the ability to use excess reagents in solid-phase synthesis (because their removal poses no problems) comes into play: even if the activated amino acid derivative is hydrolyzed, further activation can occur. While no one would advocate wanton disregard for the precautions necessary to maintain the integrity of reagents, it can be more convenient to perform reactions if rigorously inert atmospheres are unnecessary.

Classical peptide coupling methods use carbodiimides such as dicyclohexylcarbodiimide (DCC) and diisopropylcarbodiimide (DIC). If used alone, these lead to the formation of the amino acid anhydride, which is the acylating agent for an amine linked to the support. The by-product of carbodiimide activation is the urea, which can sometimes be difficult to separate from products, especially in solution-phase synthesis. A third reagent, ethyldimethylaminopropyl carbodiimide (EDC), was developed so that all products derived from the carbodiimide could be removed by an aqueous acid wash. This is an example of a principle to be discussed in the solution-phase synthesis section: acid/base properties can be used to move reagents and by-products to a different phase.

| DIC | DCC | EDC | NHS | HOBt |

Carbodiimide coupling agents and additives

Additives, including *N*-hydroxysuccinimide (NHS) and *N*-hydroxybenzotriazole (HOBt), can also be used in carbodiimide couplings. They facilitate the reaction and suppress racemization and other side reactions. Reactions promoted by NHS or HOBt proceed via formation of the so-called active esters. Couplings with these reagents have lower tendency for deprotonation at the α-carbon, the cause of racemization, than with anhydrides.

Carbodiimide coupling and coupling via an active ester

Active esters can also be obtained as stable reagents such as the Fmoc-pentafluorophenyl (Pfp) esters by treatment of the acid with pentafluorophenyl trifluoroacetate. Likewise, internal anhydrides can be prepared by treatment of the Fmoc-amino acids with phosgene, forming the *N*-carboxy anhydride (NCA) derivative.

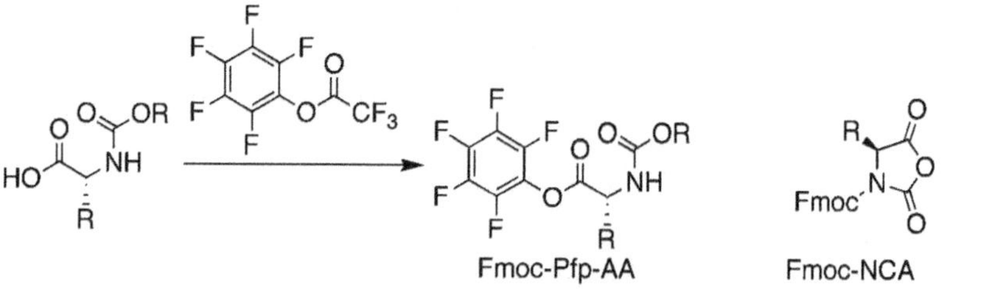

Fmoc-Pfp-AA   Fmoc-NCA

Active ester derivatives for peptide synthesis

A number of modern reagents for peptide coupling exploit phosphonium or uronium ions to generate the active ester. Hydroxybenzotriazolyltetramethyl uronium ion (HBTU) generates the Bt active ester by the mechanism shown below. The carboxylate is generated in low amounts by weak, non-nucleophilic bases such as tertiary amines (TEA, triethylamine or DIPEA, diisopropylethylamine). Phosphonium salt activators such as BOP exploit the formation of the very stable P=O bond to create the active ester by a similar mechanism.

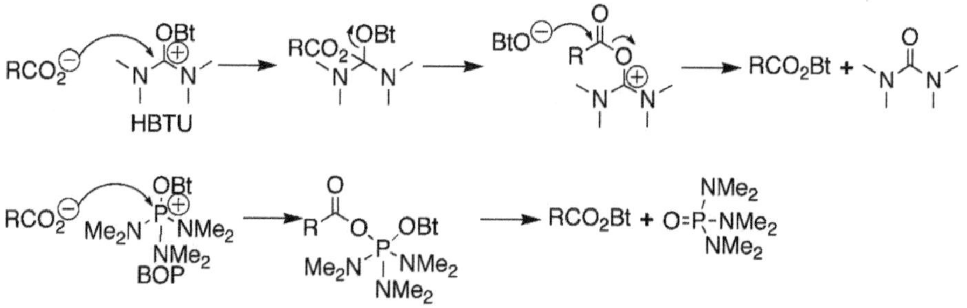

Mechanism for activation of acids by uronium and phosphonium ions

Amino acids (the 20 coded amino acids, their enantiomers, and a wide variety of unnatural amino acids) protected at their α-amino groups with either Boc or Fmoc and at side chains with appropriate protecting groups are widely available commercially. Consequently, not only have these compounds seen extensive use in creating non-natural peptides that may have resistance to proteases, they were utilized in the preparation of many early non-peptide combinatorial libraries. The commercial availability of these particular compounds stimulated the development of new, non-peptide chemistries. The ability to buy rather than needing to make reagents is probably nowhere more important than in combinatorial chemistry. The large number of reagents required to prepare a combinatorial library can make it simply impractical to use building blocks that must be prepared in multi-step syntheses.

Problems

1. The Fmoc and Boc methods of peptide synthesis require different types of supports. What main criterion distinguishes supports for each type of synthesis?

2. How would the yield of the following process be affected if the reaction was conducted on (aminomethyl)polystyrene resin (1.12 mmol/g) or TentaGel-amine resin (0.32 mmol/g)?

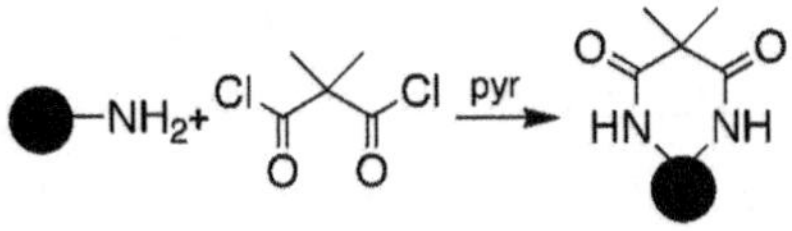

Supports and Linkers for Small Molecule Synthesis

New resins and linkers

Many novel resins and linkers have been developed to facilitate applications of solid-phase synthesis to organic compounds that would have little place in peptide chemistry. However, the existing peptide synthesis resins have certainly been used for small molecule library synthesis.

Attaching acid derivatives to supports either very robustly or very tentatively can be done with divergent approaches. The Sasrin (for super acid-sensitive resin) support is essentially Wang resin with an additional methoxy group. Acids can be removed from Sasrin with concentrations of TFA as low as 0.5 % in CH_2Cl_2.

Sasrin solid-phase synthesis support

A common functionalization of polystyrene is bromination by electrophilic aromatic substitution. The resulting bromide readily participates in metal-halogen exchange, creating a lithiated polystyrene that can be treated with a wide range of electrophiles – essentially any reagent that would react with phenyl lithium.

Triphenylmethyl (trityl) and chlorotrityl resins are derived from (lithio)polystyrene by treatment with a benzophenone. The resulting triarylcarbinol is converted to the tertiary chloride, which reacts with nucleophiles. Trityl resin is more often used for alcohols (and other less acidic functionalities) than for carboxylic acids. This demonstrates the effect of the leaving group in the cleavage process as distinguished from the effect of the carbocation. Cleavage of a

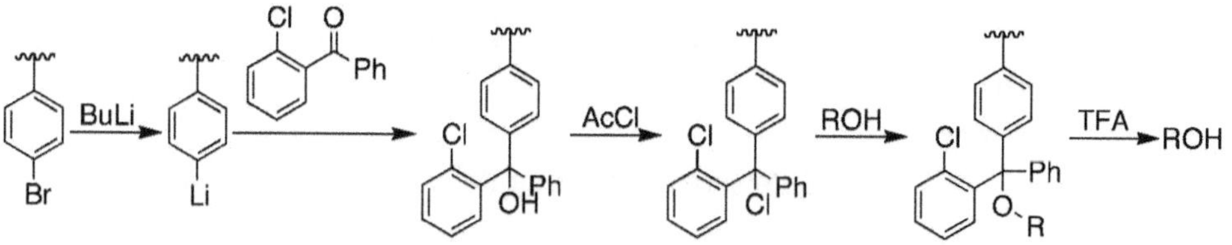

Preparation and use of trityl resins

trityl ether gives the alkoxide, but cleavage of a trityl ester gives a carboxylate that is ~10^{10} less basic. Trityl esters are thus very reactive and would likely not tolerate many reaction conditions. Based upon the inductive effect of chlorine, chlorotrityl derivatives are more resistant to acid.

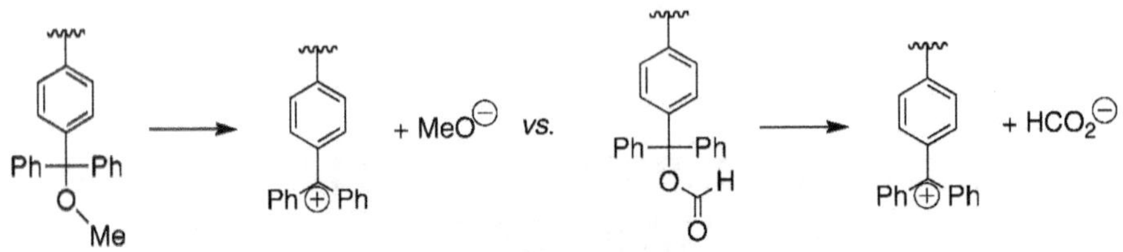

Relative reactivity of trityl resins

The protection of alcohols with trityl chlorides is Sn1-like, and the loading reaction is generally performed in the presence of weak tertiary amine bases such as pyridine, since HCl is a by-product. Cleavage from trityl ether supports can be accomplished with TFA.

The safety catch strategy is a general approach to maintaining a compound on a support through a variety of demanding reaction conditions, then in the penultimate step of the synthesis activating it for cleavage from the support. One example of a safety catch linker for carboxylic acids is the benzenesulfonamide. This known support was exploited and enhanced by Ellman. Acids can be loaded onto the support by conventional activators. The amide can be used in base-promoted α–anion chemistries (kinetic enolate alkylation, aldol condensation) without concern for loss of the link to the support. The acidity of the NH group of the sulfonamide means that it is deprotonated first, making acyl substitution reactions at the carbonyl group negligible. After the desired transformation is complete, the nitrogen is alkylated with the quite reactive alkylating agent iodoacetonitrile. This step requires only mild, non-nucleophilic base, such as diisopropylethylamine. The alkylation product is no longer protected from acyl substitution by the free NH, and in fact the inductive effect of the nitrile significantly enhances the reactivity of the carbonyl group over even the traditional methylation. Consequently, ami-

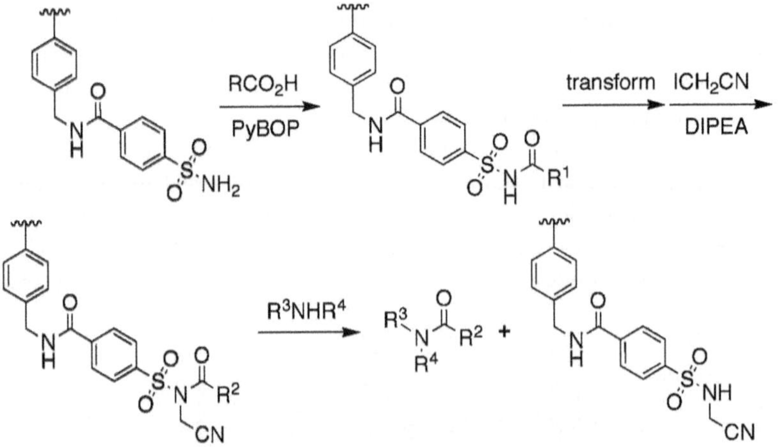

Safety catch linker strategy

nolysis of the sulfonamide can be performed with only stoichiometric amounts of amine. This enables use of non-volatile amines, similar to the advantage offered by the Kaiser oxime resin.

Two novel approaches for the linking of alcohols to supports will be described. A dihydropyran alcohol can be used to substitute for the chloride of Merrifield resin. Loading the resulting resin with an alcohol under mild acid conditions is equivalent to forming a tetrahydropyranyl (THP) ether. Removal of the alcohol from the support is accomplished with pyridinium *p*-toluenesulfonate (PPTS) in methanol, which promotes a trans-acetalization reaction. This support can also be used to load anilines (but not simple amines, as the aza-acetals are unstable - aliphatic amines are too basic). The supported anilines can be further used in organometallic reactions.

Dihydropyran resin

The second approach is based on the classic silyl protecting group. The starting resin is available by hydroboration of the allylic silane and Suzuki coupling with (bromo)polystyrene. This resin is converted by trifluoromethanesulfonic acid (triflic acid) to the silyl triflate through a protiodesilylation reaction. Silyl triflates have been well used in organic synthesis and are very reactive to substitution with almost any alcohol in the presence of weak, tertiary amine/pyridine bases. The release of the alcohol from the support is performed with excess HF/pyridine complex, which is considered quite mild, equivalent to a pH in aqueous solution of about 5. The evaporation of HF itself would be problematic, but quenching the HF with excess ethoxytrimethylsilane converts it to two inert and volatile products: ethanol and trimethylsilyl fluoride.

Silyl resin

The protiodesilylation reaction is reminiscent of a method that was earlier used in so-called "traceless linkers." An ongoing theme in solid-phase synthesis is that a functional group within the molecule to be synthesized serves as a site of attachment of that molecule to the support during synthesis. The consequence of this tradition is that most of the compounds produced by solid-phase methods will have somewhere within them an amide, carboxylic acid, amine, or alcohol. This group is a "trace" of the point of attachment. Since one goal of combinatorial chemistry is to access as diverse a collection of compounds as possible, being limited to compounds bearing these functional group limits the diversity. Thus, an attachment that is not based on one of these groups is needed. Ideally, there would be no trace of the point of attachment once the molecule is released from the support.

The reaction of aryl silanes with acid is effectively an *ipso* electrophilic aromatic substitution with hydrogen as the electrophile. Because it replaces an aryl C-Si bond with a C-H bond, protiodesilylation of an aryl silane constitutes a traceless linker. A simple example shows how the traceless linker is used. The biaryl was prepared on solid support by a Suzuki coupling process that allowed a variety of aromatic groups to be added to the resin. TFA treatment produces the free biaryl.

Protiodesilylation and its use to remove a molecule from a support

Ring-forming cleavage

Solid-phase synthesis reactions must proceed in as close to quantitative yield as possible because removal of byproducts/contaminants/impurities is not possible. Failures at intermediate stages give low yields as well as products that are impure. However, if only those compounds that successfully complete the synthesis are selectively cleaved from the support in the last step, it is possible to obtain pure products with less than perfectly efficient syntheses. Because this method most often involves formation of a ring, it has been called "cyclitive cleavage" or "cyclative cleavage." As the etymology of cyclative is questionable, we will use ring-forming cleavage. However, the term is found in the glossary of terms in combinatorial chemistry developed by IUPAC:

> Cyclative Cleavage: Cleavage resulting from intramolecular reaction at the linker which results in a cyclized product. The cleavage may also act as a purification if resin-bound side-products are incapable of cyclizing, and thus remain attached to the solid support on release of the desired material. Diketopiperazine formation … is one well-known example of cyclative cleavage.
>
> Taken from *Pure Appl. Chem.* **1999**, *71*, 2349-2365.

Examples of ring-forming cleavage reactions are shown on the facing page. The first example, a ring-closing metathesis reaction, can occur only if the amine has been successfully acylated with an unsaturated amino acid. Another ring-forming cleavage reaction (mentioned in the definition) is diketopiperazine formation, which is sometimes a side reaction during peptide synthesis. It most often occurs in the coupling to the support of the third amino acid in the peptide sequence. At that time, the group subject to cyclization is an ester, which is more reactive than the amide that is six atoms away from the free amine in all subsequent steps of a solid-phase peptide synthesis.

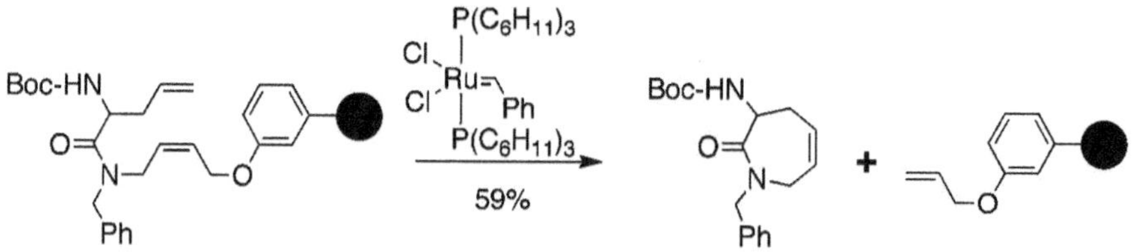

Ring-forming cleavage by the ring-closing metathesis reaction

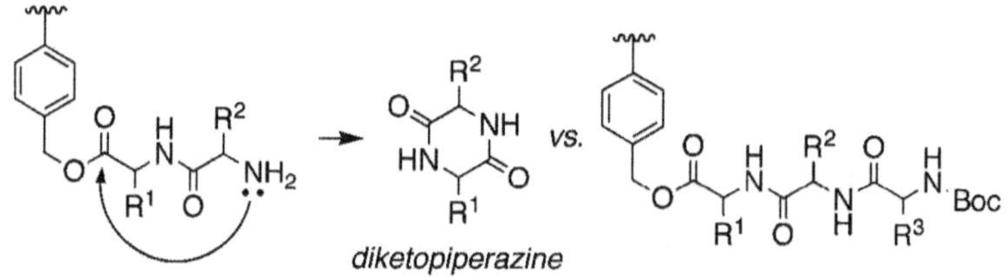

diketopiperazine

Ring-forming cleavage by diketopiperazine formation

Loading

The importance of the loading level on the resin to the outcome of a solid-phase synthesis was discussed earlier. The loading of commercially available resins is generally provided by the manufacturer and is fairly reliable. Subsequent reactions will modify (reduce) the loading of molecules of interest on the support. Reaction of benzyl alcohol with 1 g of a 1 mmol/g acyl halide resin may not proceed in quantitative yield, for example, because of competing hydrolysis of the acid chloride. Even if the reaction proceeds in 100% yield, the loading will change due to the reaction. This is because the molecular weight of the compound on the resin has increased, causing an increase in the mass of each bead, making fewer beads per unit mass. The number of desired molecules on each bead has stayed the same or decreased, meaning the number of molecules per unit mass decreases.

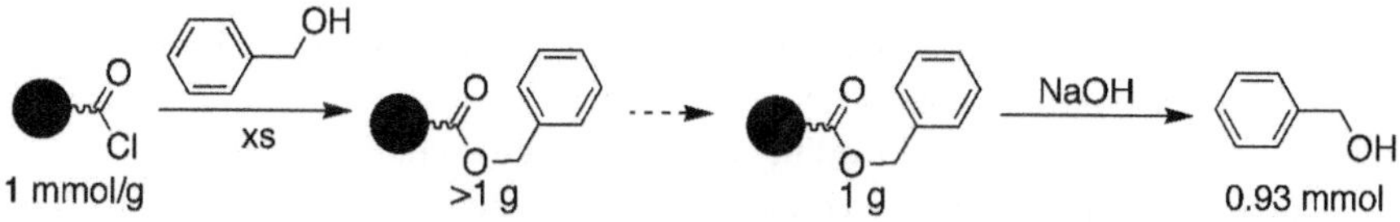

Determination of resin loading by cleavage

Resin loading is most easily determined by covalently attaching a molecule bearing the functional group of interest to a support, washing the resin extensively, and then removing that same compound from a known mass of resin. The amount of compound released can be determined by weight, or in a few cases by spectroscopic techniques. The example provided earlier of the determination of cycle yields in Fmoc peptide synthesis uses the latter technique. Another prominent example, from nucleic acid synthesis, uses the red cation generated upon removal of the dimethoxytrityl (DMTr) group, which has a 500 nm absorption.

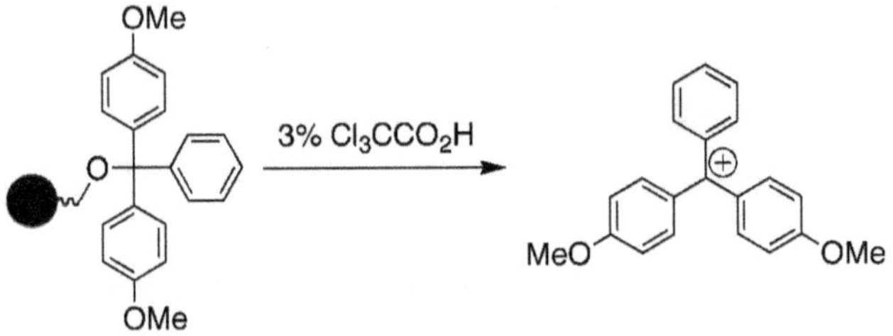

3% Cl$_3$CCO$_2$H

Determination of loading by DMTr release

Additional reading

Guillier, F.; Orain, D.; Bradley, M. Linkers and cleavage strategies in solid-phase organic synthesis and combinatorial chemistry. *Chem. Rev.* **2000**, *100*, 2091-2157.

Blaney, P.; Grigg, R.; Sridharan, V. Traceless solid-phase organic synthesis. *Chem. Rev.* **2002**, *102*, 2607-2624.

Tallarico, J. A.; Depew, K. M.; Pelish, H. E.; Westwood, N. J.; Lindsley, C. W.; Shair, M. D.; Schreiber, S. L.; Foley, M. A. An alkylsilyl-tethered, high-capacity solid support amenable to diversity-oriented synthesis for one-bead, one-stock solution chemical genetics. *J. Comb. Chem.* **2001**, *3*, 312-318.

Plunkett, M. J.; Ellman, J. A. Germanium and silicon linking strategies for traceless solid-phase synthesis. *J. Org. Chem.* **1997**, *62*, 2885-2893.

Backes, B. J.; Virgilio, A. A.; Ellman, J. A. Activation method to prepare a highly reactive acylsulfonamide "safety-catch" linker for solid-phase synthesis. *J. Am. Chem. Soc.* **1996**, *118*, 3055-3056.

Problems

1. Sasrin resin is labeled at its alcohol with heavy oxygen (^{18}O). It is coupled to Fmoc-glycine, deprotected, and acetylated. After the product is cleaved from the resin, its mass spectrum is taken. What molecular ion is observed?

2. How would you measure the loading of nonanol onto 2-chlorotrityl resin?

3. Suggest resins that could be used for solid-phase synthesis with the following building blocks. How would the compounds be put onto the resin, and what conditions would remove them?

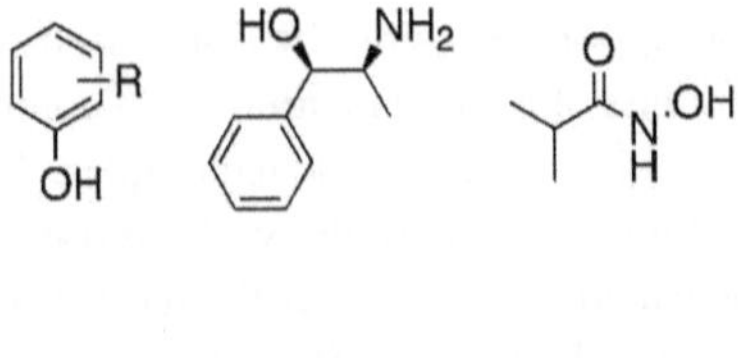

Encoded Combinatorial Chemistry

When using the split/couple/mix method to generate and test peptide libraries, molecules on the active beads are identified by very sensitive peptide sequencing techniques. As the interest in preparing libraries of molecules expanded outside of peptides, a problem arose. If the molecules to be made by the split/couple/mix method are not peptides, how are the structures of the active compounds to be determined at the level of a single bead?

Two general solutions to this problem have emerged. One is self-encoding: to directly determine the structure of the compounds themselves using physical methods, primarily mass spectrometry. MS is the only technique with sufficient sensitivity to analyze the amount of compound on a single bead (though sometimes the larger beads with higher loading must be used). For example, 500 μm polystyrene beads can carry ~150 nmoles of compound. This is sufficient to obtain mass spectra of compounds removed from these supports. It might seem wildly optimistic to think that the structure of an unknown compound could be determined from only a mass spectrum, but given that the components that have gone into the library are known and the molecular weights of every possible product can be calculated (provided that the synthesis gives the expected target), this is not such a difficult problem. Even if some library components have the same molecular mass (are *isobaric*), the probability is low that more than a few isobaric compounds would be among the interesting actives, and if need be each could be made and directly re-tested. The second method to deal with this problem is to use a molecular code to directly describe the structure of the compound that was prepared on each bead.

The concept of molecular encoding goes all the way back to the central dogma of molecular biology itself, where DNA encodes protein. Closer to the combinatorial chemistry world, peptide libraries on phage, an advanced subject covered later, use DNA to encode the sequence of peptides in a library and to read out the structure when an active is identified.

A readily sequencable set of molecular "codes" or "tags" on each synthesis bead that tells its synthetic history provides a method to directly determine molecular structure on a very small scale. Codes have included native peptides, DNA, secondary amines, and isotopically-substituted compounds. The codes have been read by Edman degradation, polymerase chain reaction (PCR)/DNA sequencing, fluorescence-detected HPLC, electron-capture gas chromatography, and mass spectrometry. All of these methods were chosen for their high sensitivity (pmol or better), because they must be able to decode single beads, and the amount of tag on a bead can be as low as 1 pmol.

During the preparation of an encoded library, chemistry is being performed to create the molecules with the activity, while chemistry is also being performed to attach the codes. The encoding chemistry is additional "overhead" in a synthesis that can already be fairly complicated, so the resulting library generally must be quite large to justify the extra effort.

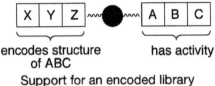

encodes structure   has activity
of ABC
Support for an encoded library

The encoding of the library is ingenious, and draws on analogy to the base-4 DNA code. It generally is a binary code where the presence or absence of X, Y, Z, ..., corresponds to one bit in an n-digit binary number. It is these binary numbers that identify the different building blocks A, B, C. This strategy is contrasted with an alternative in which A, B, C are directly represented by the presence or absence of X, Y, Z. The virtue of binary encoding is that it is also combinatorial. If each building block were encoded by one tag, as many tags as building blocks would be required. With n codes, the binary strategy can encode $2^n - 1$ building blocks. While there are 2^n combinations of n bits, one combination has all bits set to 0. Because the absence of a signal for any of the codes X, Y, and Z could be due to any type of failure in the chemical system (a negative experiment), it is unwise to use this all-null code word to represent an actual molecule.

The size of the library that can be encoded with a set of molecular tags can be calculated using the formula given above. This calculation shows that

$$2^{20} - 1 \approx 10^6,$$

therefore a set of 20 chemical tags is sufficient to encode a library of quite significant size.

One of the most widely known tagging systems was developed by Still. The tags are halogenated aryl ethers bearing alcohol chains of varying length. This tagging method takes advantage of analytical techniques used to detect haloaromatic compounds at very low levels, which are a result of the significant environmental interest in these compounds.

One technique for analysis of haloaromatics is electron-capture GC (EC-GC). Separation of compounds is based on conventional gas chromatography, but the detection is specific for those compounds with a low reduction potential, such as haloaromatics. As compounds flow past an electrode held at a controlled potential, those whose reduction potential is below that of the electrode potential will capture an electron from the electrode, and that electron transfer can be detected. The reduction potential is related to the stability of the resulting radical anion, hence the potential is strongly influenced by aromatic substitution (like halogens). The sensitivity of the technique is dependent on the reduction potential of the aromatic, but can be as low as femtomoles. EC-GC provides a means to read codes based on their different GC

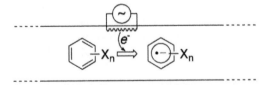

Electron–capture gas chromatography detection

retention times, which are achieved simply by changing the length of an alkyl chain. The tags are linked onto a support bead during the synthesis by an α-ketocarbene insertion process. The link between the support and the tag includes a catechol bis-ether. Ethers are generally quite stable, but catechol ethers are sensitive to strong oxidants such as ceric ammonium nitrate (CAN). Oxidation by successive one-electron steps gives an o-quinone hemiacetal that readily hydrolyzes to the alcohol, which is linked to the haloaryl ether. To enhance the GC properties of the resulting alcohols, they are converted to their trimethylsilyl ethers before analysis.

Tags must be inert to many different types of chemistry, and these Still tags are among the most robust. Two main design criteria for tag inertness are: tags must remain structurally invariant and attached to the support throughout the synthesis, yet be readily released for decoding; tags must have no functionalities that interfere with the chemistry that will be used in creating the library members.

A tag and the release of the haloaryl ether from the support for EC-GC analysis

Regarding the first criterion, the attachment of the tag to the bead can be very robust because the reaction conditions for tag release can be quite stringent. This is because decoding generally is performed after testing of the synthetic library members, so it is unnecessary that tag release be harmless to them. To make the tagging method widely applicable, the chemistry used for tag release should be used relatively rarely in organic synthesis, minimizing the likelihood of interference with library construction chemistry. The Still tags obviously could not tolerate strongly reducing or strongly oxidizing conditions, but they are stable to a wide variety of other acid- and base-promoted chemical reactions. Regarding the second criterion, tags must also be chemically unreactive to a wide variety of conditions so they do not interfere with any of the reactions that might be used in library creation. The aryl ethers and haloaromatics of the Still tags fit this bill as well.

Another feature of this tagging system as practiced at one company, Pharmacopeia, is the photochemical release of library compounds for assay in solution. Solution-based assays are the most versatile, but could lose the hard-won information captured by encoding. This potential problem is addressed by spatial identification of pools of beads. The photolabile linker enables short duration photolysis to release only part of the compound on a bead into solution for testing. This makes repeated testing of each bead possible. The beads are distributed in pools, 20 beads per well, into 96-well microtiter plates. Photolysis and testing of the filtrate from each well identifies active wells. If a well is active, its beads can be distributed as singles into individual wells, more compound can be removed by photolysis, and testing can be repeated. This protocol limits the amount of decoding in the analysis of a library, since only those beads that have produced active single compounds are decoded.

Partial photochemical release and testing

An example will serve to illustrate how this system operates in practice. The goal is to find inhibitors for the enzyme carbonic anhydrase, which is a drug target for diseases such as glaucoma. Sulfonamides are known to strongly inhibit carbonic anhydrase, so that particular functionality will be included as part of the library. Other diverse structures would be included in the library that might impart specificity for a particular type of carbonic anhydrase. The molecular tags are diazoketones linked through an aromatic ring and α,ω-diols to trichlorophenyl and pentachlorophenyl ethers. In the scheme, the codes are denoted by "primed" R groups, while the chemical diversity of the library is denoted by superscripted R groups.

The library synthesis begins with beads bearing the photocleavable linker. They are split into 7 pools and one of 7 uncommon amino acid or alcohol building blocks is added to each. Each building block includes amino groups protected with Fmoc. One tag molecule is chosen to represent each of the three bits in a code for this first diversity element, and each pool is encoded with a particular combination of 1-3 molecular tags. The 7 combinations that can be encoded by 3 molecules are depicted in brackets. The encoding reaction is a carbenoid insertion into (primarily) the polystyrene support promoted by rhodium trifluoroacetate. The tags are incorporated at less than the 5% level.

After encoding, these bead pools are combined and split into 31 pools. The Fmoc group is removed and 31 Fmoc amino acids are coupled to the liberated amino groups using standard peptide chemistry methods. Encoding of those pools is then performed. New tag molecules are chosen to represent each of the five bits in a code for the second diversity element, and each pool is encoded with a particular combination of 1-5 molecular tags. The brackets here reflect the 31 combinations that can be encoded by 5 molecules.

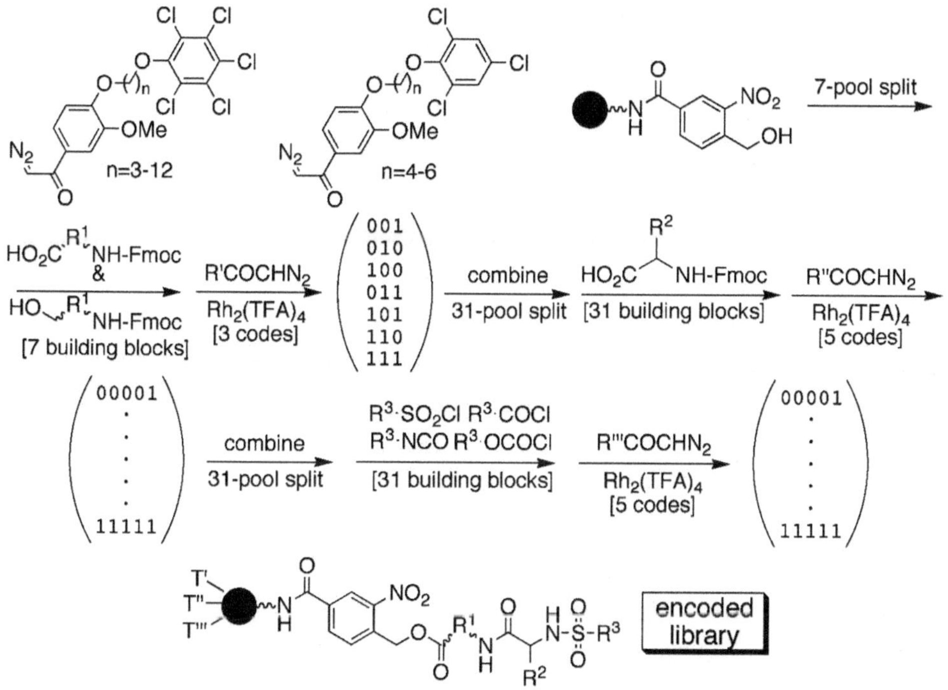

Synthesis of a carbonic anhydrase inhibitor library

After this second encoding, the bead pools are again combined and split into 31 pools. The Fmoc group is removed and one of 31 acylating or sulfonylating reagents is added. These 31 bead pools could be kept separate for compound release from the support and testing. The 217 compounds from each pool could be assayed together, while recording which reagent was used for acylation or sulfonylation in each. The best well would presumably identify the best third diversity element. This approach would suffer the difficulties of pools earlier discussed, however. In this case, encoding of the third position pools was performed. Another set of tag molecules was chosen to represent each of the five bits in a code for the third diversity element, and each pool was encoded with a particular combination of 1-5 molecular tags. These pools could be combined to complete the preparation of a library of 6727 compounds with each compound's identity defined by up to 13 molecular tags.

Though it might appear at first that the choice of the number of pools to be used in each splitting step and therefore the number of building blocks incorporated into the library is random, it is not. The numbers 7 and 31 are solutions of $2^n - 1$ for n = 3 and 5. For a given number of molecular tags used to represent a diversity set, this value is the maximum number of building blocks that could be encoded. Five molecular tags would also be required to encode 30 building blocks, or 20 building blocks, or in fact any number of building blocks greater than 15, the value of $2^n - 1$ for n = 4. Therefore, the maximum use is made of the available diversity in encoding molecules when $2^n - 1$ building blocks are encoded in n codes. Unlike the genetic code, which has 64 combinations encoding 20 amino acids, such a code is not degenerate.

The completion of this library experiment involves the screening process described earlier. Beads are portioned into wells 20 at a time and compounds are released photochemically for screening in solution. The wells that contain active beads are then split into 20 single-bead wells. After a second photochemical release, elution from the bead gives single pure compounds for testing. When an active well is discovered in this step, it is possible to go back to the well from which the active was eluted and isolate the bead from which it came. That bead is subjected to the decoding procedure described earlier to provide the structure of the active product.

active bead

Testing and decoding of a carbonic anhydrase inhibitor library

Where is the efficiency in this process if individual compounds are ultimately tested? In the synthesis: 7 + 31 + 31 steps are used to prepare 7 × 31 × 31 compounds. However, there is some inefficiency: tagging chemistry adds "overhead" in addition to the synthesis chemistry needed to prepare the targets. However, all of the tags can be added simultaneously in one step, using up to 5 tags. Situations can also be arranged in which the tags have the same chemistry as the building blocks and can therefore be added at the same time as the main diversity elements by "doping" them into these reagents.

Many alternative encoding systems have been developed. Eliminating the step in which the tag is released simplifies tag synthesis and use, but requires that tags are decoded while still on the support. Spectroscopic methods are usually required for this task. For example, infrared spectroscopy (IR) techniques can be used to read tags consisting of alkynes and nitriles. Dye-impregnated silica colloids can be used as tags and read by fluorescence microscopy.

The Still tagging system is currently used for encoding of small molecule combinatorial libraries at the Institute of Chemistry and Cell Biology (ICCB) at Harvard and in industry at Pharmacopeia.

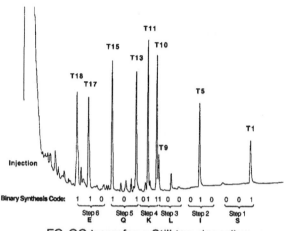

EC-GC trace for a Still tag decoding

Reprinted with permission from *Proc. Natl Acad. Sci. USA* **1993**, *90,* 10922-10926.

Additional reading

Still, W. C. Discovery of sequence-selective peptide binding by synthetic receptors using encoded combinatorial libraries. *Acc. Chem. Res.* **1996**, *29*, 155-163.

Burbaum, J. J.; Ohlmeyer, M. H.; Reader, J. C.; Henderson, I.; Dillard, L. W.; Li, G.; Randle, T. L.; Sigal, N. H.; Chelsky, D.; Baldwin, J. J. A paradigm for drug discovery employing encoded combinatorial libraries. *Proc. Natl. Acad. Sci. USA.* **1995**, *92*, 6027-31.

Problems

1. Using the fluorinated compounds shown below, which each have different chemical shifts for ^{19}F NMR decoding, how many different building blocks could be encoded in a single diversity step of a binary encoded combinatorial library?

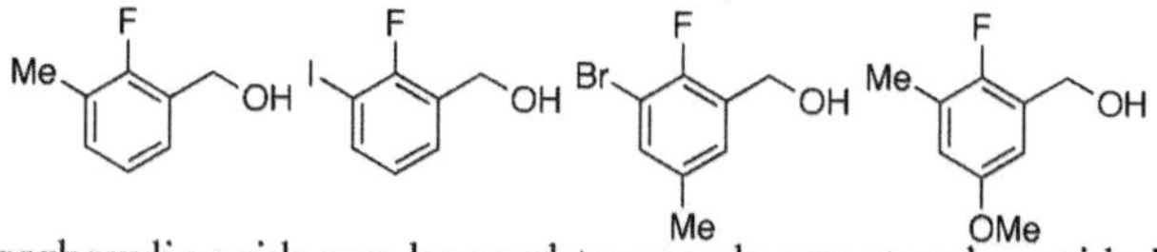

2. Triazine carboxylic acids can be used to encode unnatural peptide libraries by addition to the resin at the same time as activated amino acids. When the triazine is encoded by isotopes of nitrogen and hydrogen, decoding may be performed by mass spectrometry following the removal of triazine from the bead by methanolysis. Using the two most common isotopes of each element, how many unique mass codes can be created from triazine carboxylic acid? If a dipeptide library is created using this method, how many unnatural amino acids could be encoded with these tags?

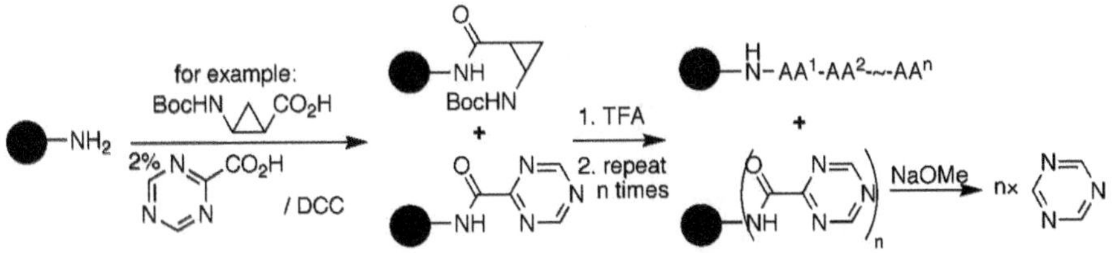

3. In one variant of the Pictet-Spengler reaction, an aldehyde is condensed with a phenethyl amine to give the corresponding tetrahydroisoquinoline. The extension of this reaction to solid support has been accomplished.

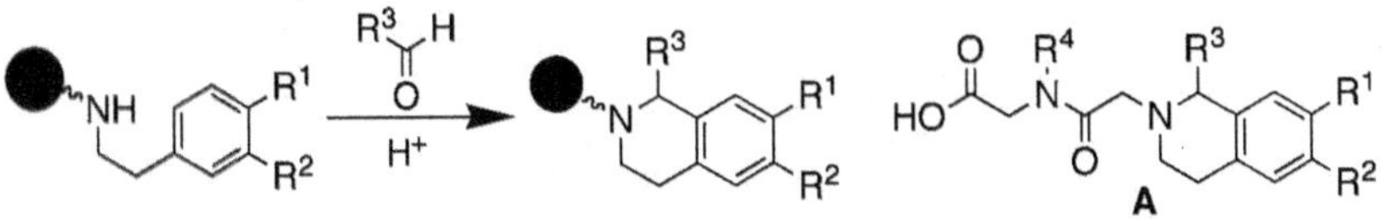

Design a split-mix synthesis of a minimum of 600 compounds of the general structure **A** using 3–5 different tyramines (R^1, R^2), 10–20 different aldehydes (R^3), and unlimited primary amines (R^4). Use the Still encoding method, accounting for it explicitly in your scheme. Describe how you would test the library given an assay in 96-well plate format. How would you identify the positive compounds?

Directed Sorting

Another combinatorial chemistry method appears on its surface to be an encoding technique, but on closer inspection directed sorting falls into a different class. In fact, the closest precedent/analogy for this work is the Houghten tea bag method for peptide synthesis. Recall that this method involves SPPS beads that are kept physically isolated in mesh packets. Several different packets can be treated batch-wise for the common steps of peptide synthesis and individually for coupling of different amino acids. The tea bag method was not covered in detail earlier, but this discussion will present similar concepts.

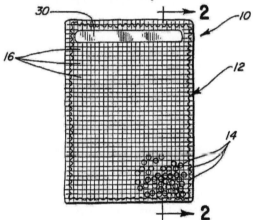

Schematic of "tea bag" from Houghten patent US 4,631,211

Two companies, Ontogen and IRORI, developed similar methods for directed sorting. IRORI has made their technology broadly available, and most of the publicly available information on the method comes from IRORI (now Discovery Partners). IRORI uses an integrated circuit device called an RF (radio frequency) transponder that is similar to the RFID devices used to collect tolls on freeways and bridges. The transponder receives RF energy from a reader that is then used to transmit back to the reader a digital signature that is unique to that particular item. These RF transponders have been widely used for identifying laboratory animals, for example, so science applications of this high technology are not new. The microchip is housed in a 5 mm diameter glass envelope, so the transponder is inert to most chemical treatments.

RF transponder – "AccuTag" - photo

Two common misperceptions exist about the transponder technology. One is the transponder size. If transponders were truly as small as a single solid-phase synthesis bead, this would be a very powerful system that essentially would supersede all other encoding methods. However, given their macroscopic size, transponders cannot encode single beads in the same way that molecular tags can. They are instead used as electronically readable labels for porous reaction containers (called "MicroKans") for solid-phase synthesis resin. These reactors hold milligram quantities of beads and one transponder. The second misconception is that the transponder is a read/write memory that gets changed with each chemistry step. This would be *necessary* if transponders were used in a conventional split/couple/mix procedure; indeed, that is how libraries are ordinarily encoded. However, because the digital signatures are instantly and non-destructively readable, a Kan holding its individual transponder can be sorted (with its accompanying compounds), rather than pooled and split. Writing data to the transponder is unnecessary, and sorting permits the preparation of exactly the same number of compounds as the number of Kans. Directed sorting is not subject to the Poisson statistics of one-bead, one-compound libraries, so Kans are not needed in excess. These are one-Kan, one-compound libraries. Because sorting is performed by a robot, a specific transponder (based only on its unique identity) can be placed into a specific reaction to receive a particular building block in an individual step. Hardware is available that sorts the MicroKans for a 100 × 100 synthesis (10000 compounds) overnight.

The directed sorting procedure is performed as follows. Assume we wish to prepare a 2-residue combinatorial library from the building blocks {A,B,C} in step 1 and {D,E,F} in step 2. We require nine Kans to carry the nine compounds. They are sorted into three groups of three, and each group is coupled with one pure building block. While the figure shows that these are "combined", this is not actually necessary, or even correct, as each Kan maintains its discrete existence throughout the process. The important feature is that the process is not dependent on statistics to distribute reactants to the appropriate building blocks. It specifically takes one of the Kans that was in reaction A in the first round to reaction D in the second round, one to reaction E, and one to reaction F. A similar process is applied to the product of the first round B and C reactions. The alert reader will note that this description is quite similar to the earlier discussion of an ideal split/couple/mix library synthesis. An animation can be currently viewed on the IRORI web site at the URL given on the facing page.

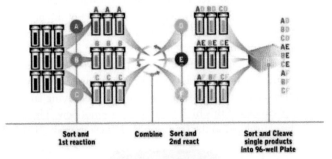

| Sort and
1st reaction | Combine | Sort and
2nd react | Sort and Cleave
single products
into 96-well Plate |

Directed sorting creates all possible combinations without reliance on statistics

http://www.discoverypartners.com/Products/irori_tech_demo_sorting.html.

One of the virtues of this transponder method, which houses the compound identifier and the support in a container, is that it uses conventional solid-phase synthesis resins. Even if one could make a transponder that would fit inside a single support bead, these would have to be functionalized with each and every handle, polymer, etc., that a user might ever need. In contrast, with Kans, it is not necessary to specially adapt to directed sorting an existing method for the synthesis of a particular target. As improvements are made in supports, these can be directly incorporated into the practice of the synthesis.

A simple example of the use of the MicroKan method for the synthesis of a 432-member combinatorial library is given below. The Rink amide resin is used in an initial reductive amination with 18 aromatic aldehydes. The product is acylated with cyanoacetic acid, and a Knövenagel reaction is performed with a set of 8 phenolic aldehydes. The products are acylated with 3 acid chlorides and the aryl esters removed from the support with TFA. The resulting compounds are tyrphostins, a family of compounds with the ability to inhibit protein tyrosine phosphatases. These enzymes are involved in several important biological signaling pathways.

Solid-phase synthesis of tyrphostin analogues

The MicroKan technology has been adopted in industry, as it offers the potential to prepare 10000 pure compounds in 6 weeks once the chemistry of a particular library has been established. However, it was impractical with this technology to increase the library size by another order of magnitude. A method to make 100000 member libraries in 10 days was desired.

With the principle of automated, directed sorting established, further extensions can be made. Non-electronic methods could be used to mark containers of beads for automated sorting, in the same way that bar codes are used to mark items in a market. In fact, optical methods of identification of the compounds during a synthesis are quite powerful, as their readout is non-destructive, and conversion of optical to digital signals is timely and straightforward. This situation contrasts with chemical encoding technologies, which require steps and time for decoding.

The IRORI approach to optical encoding is called the NanoKan. These porous plastic cylinders are about the size of a pencil eraser (6 mm; they fit snugly into a well of 96-well plate) and hold about 8 mg of support. This should lead to about 1.6 mg of compound with average molecular weight and loading. The top of each NanoKan is laser-etched with a unique pattern that can be read by the optical system that performs the sorting.

NanoKan with optical barcode

These devices were used in the synthesis of a 10000 compound screening library. The sorting of 10000 NanoKans required one day for each diversity-introducing step. The release of the compounds from the support was performed with TFA in dichloromethane in a unique Clevap plate that exploits the NanoKan size. They are pressed into the plate and incubated with cleaving reagent. The plates are centrifuged to elute the solvent into 96-well plates, which are evaporated.

Kan reactors with a wide range of capacities are currently available commercially.

Kan reactors

Additional reading

Herpin, T. F.; Van Kirk, K. G.; Salvino, J. M.; Yu, S. T.; Labaudinière, R. F. Synthesis of a 10000-member 1,5-benzodiazepine-2-one library by the directed sorting method. *J. Comb. Chem.* **2000**, *2*, 513-521.

Nicolaou, K. C.; Pfefferkorn, J. A.; Mitchell, H. J.; Roecker, A. J.; Barluenga, S.; Cao, G.-Q.; Affleck, R. L.; Lillig, J. E. Natural product-like combinatorial libraries based on privileged structures. 2. Construction of a 10 000-membered benzopyran library by directed split-and-pool chemistry using NanoKans and optical encoding. *J. Am. Chem. Soc.* **2000**, *122*, 9954-9967.

Shi, S.; Xiao, X. Y.; Czarnik, A. W. A combinatorial synthesis of tyrphostins via the "directed sorting" method. *Biotechnol. Bioeng.* **1998**, *61*, 7-12.

Problems

1. Propose a reaction scheme for the preparation of a library of salicyl hydroxamate ethers in NanoKans with the indicated points of diversity. With 24 R^1s, 16 R^2s and 24 R^3s, how many NanoKans will be present in each reaction solution in each diversity-introducing step in the synthesis?

2. A MicroKan is filled with 30 mg of a methyl sulfoxide resin with a loading of 1.2 mmol/g. The reaction sequence shown is executed for one encoded MicroKan, the product is cleaved from the resin with TFA, it is eluted with CH_2Cl_2 and the solvent is evaporated. What is the theoretical yield (in mg)?

3. How many different codes are possible for NanoKans based on a 12 × 12 laser-etched grid?

Unnatural Oligomers for Library Synthesis

The techniques originally developed to prepare libraries of peptides were clearly very powerful, but peptides do not offer optimum properties in drug discovery, as discussed earlier. A significant effort was put toward developing sets of alternative oligomeric molecules that could be assembled by sequential synthesis methods but would not be recognized by proteases. In some ways, this research also triggered interest in unnatural oligomeric molecules generally, and research into these materials (called "foldamers") and their conformational properties has been a rich area of chemical research.

Peptoids

The term peptoids was used to describe *N*-substituted glycine (NSG) oligomers in which the diversity arises from different *N*-alkyl groups. Peptoids are assembled somewhat differently than peptides. As the glycine portion of each residue is invariant, bromoacetic acid can be used as the sole precursor to this section. The diversity is incorporated by displacement of the reactive α-bromoamide by a primary amine, a reaction that is conducted overnight in DMSO. Because the repeating, monomeric unit of a peptoid is composed of both the α-bromoacid and the primary amine, this approach is referred to as a sub-monomer synthesis because the two sub-units of the monomer are added separately. In the example given below, a tripeptoid is assembled with an acylating agent on the *N*-terminus.

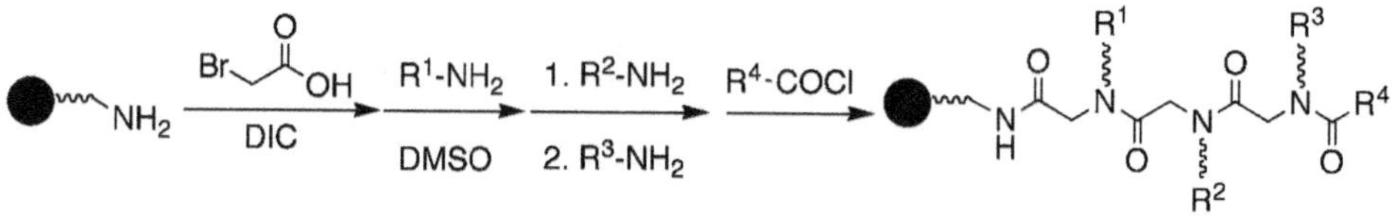

The sub-monomer strategy for peptoid synthesis

Secondary amines can be used in peptoids provided they are diamines, which allow the chain to be extended. These sub-monomers enhance diversity.

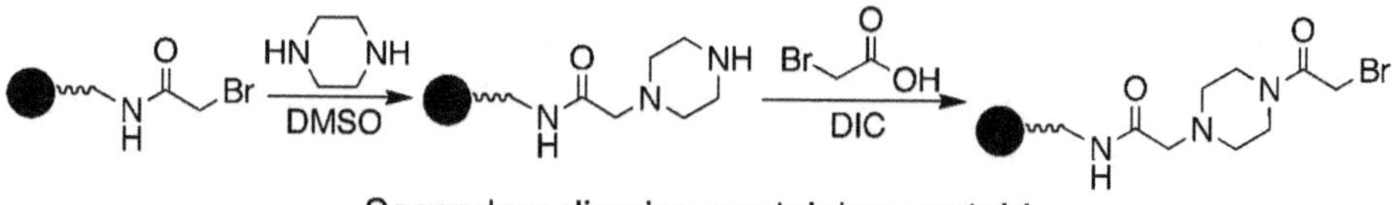

Secondary diamine-containing peptoid

These methods were used in the preparation of a screening library for the class of receptors called the seven transmembrane (7TM) helix/G-protein coupled receptors (GPCRs). Building blocks were chosen based on already-known ligands for these receptors, such as tyramine. Split/couple/mix synthesis was used to assure equimolar amounts of each peptoid in pools consisting of 204 trimers. Additionally, "null" building blocks were used in this work. One of the pools resulting from each split is purposely not coupled with any building block. First, this strategy creates all possible deletion sequences, providing an internal comparison in case one of the active sequences is not in fact the intended sequence but a synthesis failure, its deletion. Nulls also create greater diversity in the library. In this case, the null building block results in pools that contain dipeptoids in addition to the 204 tripeptoids. These peptoids were acetylated or carbamoylated at their *N*-terminus as well. Iterative testing, synthesis and deconvolution was applied to the 18 pools in this library to discover a potent ligand for the α1-adrenergic receptor.

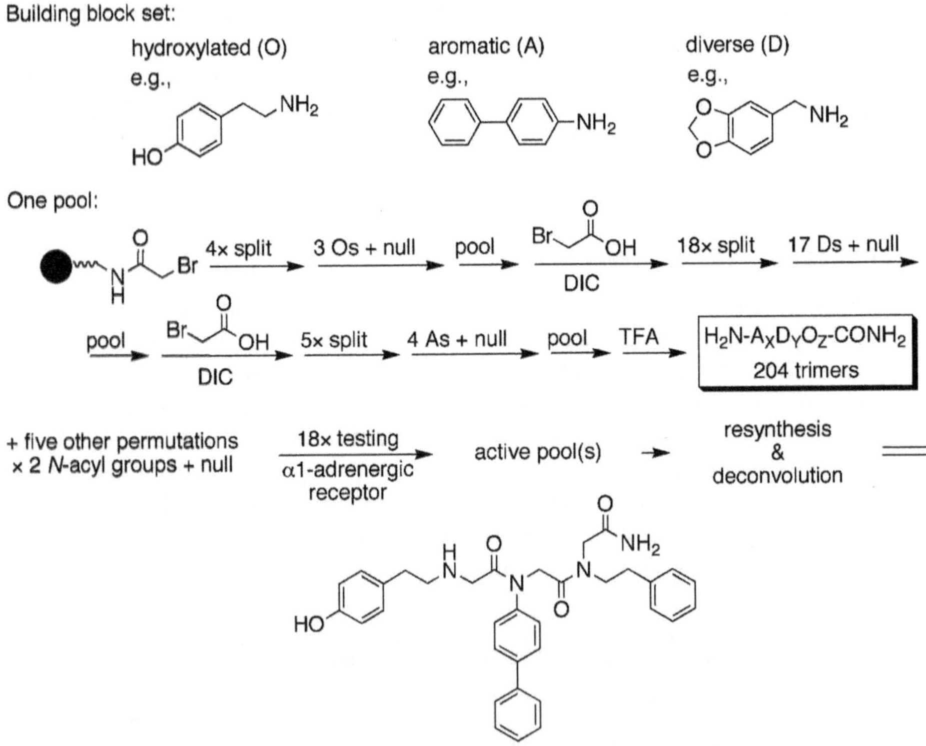

Building block set:

One pool:

Preparation of a peptoid combinatorial library and solution-based pool testing

Azatides

These oligomers are simpler even than peptoids. The glycine CH_2 of the peptoid is substituted by a nitrogen atom to make azatides, which are poly(acylhydrazines). Azatides are composed from two sub-monomers, a carbon dioxide equivalent and a family of monoalkylhydrazines. Azatides are not stable with a free carboxy terminus, as the carbazic acid will decarboxylate. In the synthesis of azatides, the position of acylation (between the two nitrogens of the monoalkylhydrazine) is controlled by a protecting group. A simple azatide was prepared as shown below. Two azatide residues replace the two glycines in the YGGFL peptide sequence that is recognized by a monoclonal antibody.

Boc-protected methyl hydrazine is used in a standard peptide coupling with Fmoc-Tyr. After the Boc group is removed with acid, an azatide glycine equivalent is used to acylate the methylated nitrogen. The acylated nitrogen does not react in this step because it is deactivated electronically. The next Boc group is removed and the unacylated nitrogen adds to the isocyanate derived by treatment of a protected FL dipeptide with phosgene (Cl$_2$CO). Deprotection gives the azapentapeptide, which surprisingly did not mimic the parent peptide in antibody recognition. The azatides are thus likely to be an interesting sidelight of efforts in unnatural oligomer synthesis.

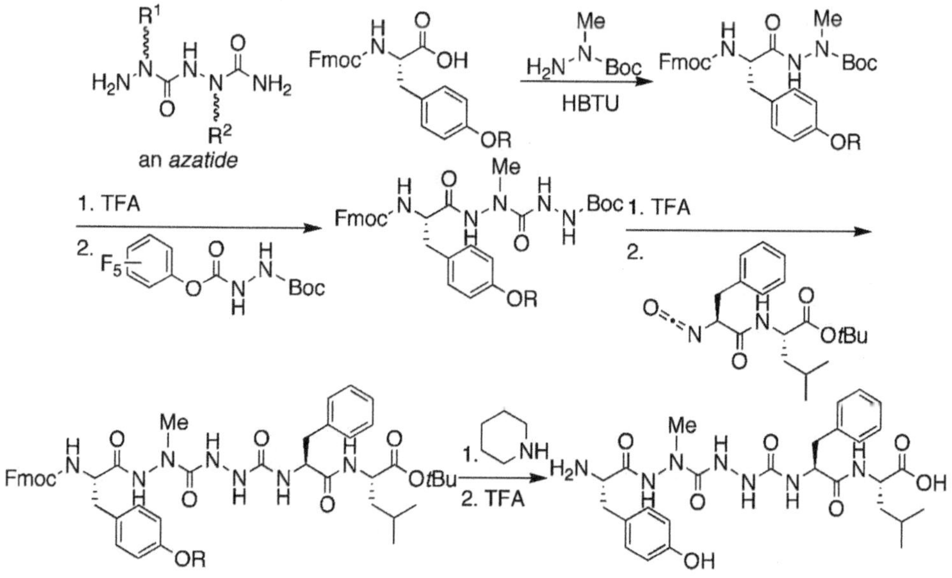

Synthesis of an azatide, YA^{a}G^{a}FL

Peptidyl phosphonates

Incorporation of phosphonamide and phosphonate units into peptide backbones has been a valuable strategy to target peptide-hydrolyzing enzymes, as the tetrahedral phosphorous serves as a good mimic of the tetrahedral intermediate in the hydrolysis reaction. These "transition state analogues" can be highly potent inhibitors of proteases. SPPS is generally not used to prepare peptide sequences that include phosphonates. Novel methodology was therefore needed to prepare a library of peptidyl phosphonates.

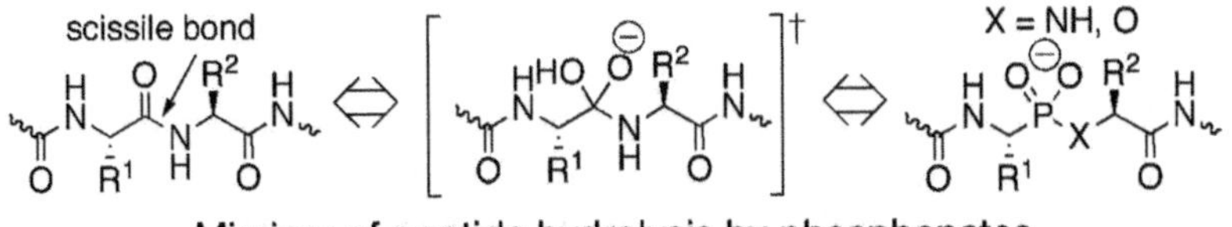

Mimicry of peptide hydrolysis by phosphonates

The α-hydroxyacids corresponding to several of the genetically coded α-amino acids are readily available and can be derivatized with an Fmoc O-protecting group for solid-phase synthesis. After deprotection, esterification of the alcohol with a protected α-aminophosphonic acid is accomplished by Sn2 displacement via the Mitsunobu reaction. Continuation of conventional synthesis creates a peptide with an embedded O-phosphonate. This method was applied to the synthesis of a 540 compound tripeptide library. The P diversity element represents 6 carbobenzyloxy (Cbz or Z) protected α-aminophosphonic acids. The O diversity element represents 5 Fmoc-protected hydroxy acids. The C diversity element represents 18 Fmoc-protected amino acids.

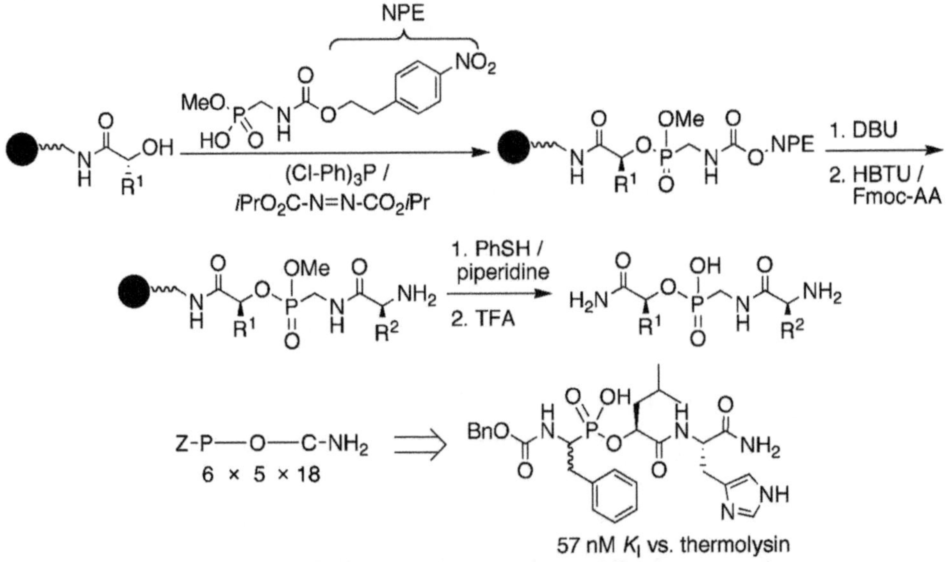

57 nM K_i vs. thermolysin

Synthesis of peptidyl phosphonates and library screening

Screening of this library was performed while the phosphonopeptides were attached to TentaGel resin. An activity depletion assay was used in which resin beads bearing phosphonopeptides are incubated with a solution of protease. Those that strongly bind the protease remove it from wells when the beads are removed. The residual activity in each well is then detected with a conventional chromogenic enzyme assay (that is, one that produces a color that can be related to the amount of product formation). Iterative deconvolution of the active compounds in this library led to the discovery of potent inhibitors of the protease thermolysin, such as the ZF[PO]LH-NH$_2$ sequence shown.

Oligoureas

These structures have much less direct correspondence to the peptide backbone than those that have so far been described. Whereas a peptide encompasses three atoms per monomer unit, the oligoureas encompass five atoms per monomer unit. They are best thought of not as peptide mimetics but as novel oligomers in their own right.

One approach to oligoureas exploits the natural amino acids as both the source of chirality and the backbone of a differentially functionalized ethylenediamine building block. One nitrogen is activated as a *p*-nitrophenylcarbamate, while the other is protected as an azide. Coupling of the carbamate to an amine on the support is followed by "deprotection" − reduction of the azide to the amine by stannous chloride sets up the next synthesis cycle. The oligoureas were investigated as replacements for different residues in the YGGFL peptide, and the sequence shown was discovered.

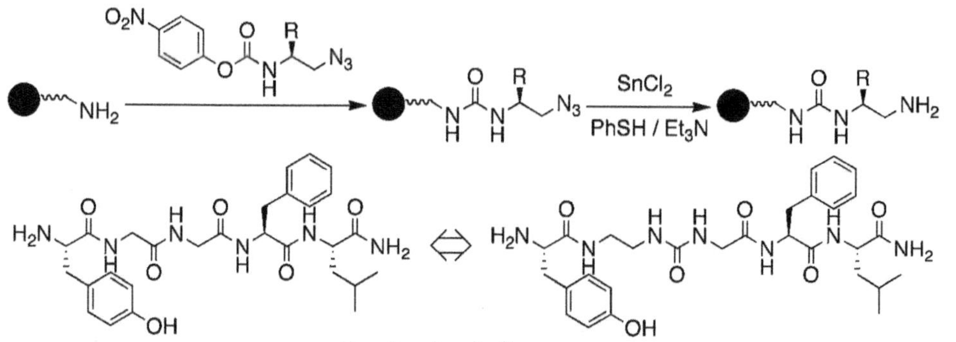

Synthesis of oligoureas

Additional reading

Figliozzi, G. M.; Goldsmith, R.; Ng, S. C.; Banville, S. C.; Zuckermann, R. N. Synthesis of *N*-substituted glycine peptoid libraries. *Meth. Enzymol.* 267, Academic Press; San Diego (1996) p. 437-447.

Janda, K. D.; Han, H. Azatides: Solution and liquid phase syntheses of a new peptidomimetic. *J. Am. Chem. Soc.* **1996**, *118*, 2539-2544.

Burgess, K.; Ibarzo, J.; Linthicum, D. S.; Russell, D. H.; Shin, H.; Shitangkoon, A.; Totani, R.; Zhang, A. J. Solid phase syntheses of oligoureas, *J. Am. Chem. Soc.* **1997**, *119*, 1556-1564.

Problems

1. The complete set of anilines with two aromatic *C*-methyl groups is used for the building blocks in a peptoid synthesis. Describe the operations involved in one cycle of the synthesis using the split/couple/mix method.

2. Suggest a molecule that might inhibit proteases that cleave proteins specifically at Phe-Ala sequences. Explain.

Synthesis of oligoureas

Analytical Methods for Solid-phase Synthesis

Determining the outcome of a reaction that has been performed on a solid phase is a challenging prospect. Two issues that arise in any chemical reaction, product identity and yield, must be addressed. The methods used to determine these outcomes are commonly different.

Product identification

The development of synthetic chemistries on the solid phase faces difficulties that do not concern conventional chemistry. Many common analytical techniques, such as thin-layer chromatography, are unavailable, at least directly. It is always possible to remove reaction intermediates from the resin to subject them to various analytical techniques, but in fact the time and trouble required to do this most often means that it is not done. Even non-destructive analytical methods are effectively destructive when they require the compound to be removed from the bead for analysis. With one-bead, one-compound libraries, cleavage of the compound for analysis would destroy the hard-won uniqueness of each bead, and the sheer number of beads/compounds could swamp conventional analytical techniques when performed serially. The amounts of compound on each bead can be small to begin with, so only the most sensitive analytical techniques, such as mass spectrometry, can be applied to analysis of cleaved compounds. Consequently, many methods have been developed to directly follow reactions on the solid phase.

Gel-phase NMR

Surprisingly good NMR spectra can be obtained from compounds bound to polystyrene beads, which exist in a gel phase when swollen in appropriate solvents. ^{1}H NMR spectroscopy is generally not useful, but nuclei with a large chemical shift dispersion and limited or no splitting can be used to follow reactions on polymer supports. Broad spectral line widths are commonly observed for polymers because of their low mobility in solution and relatively long relaxation time. Molecules attached to polymer beads by a linker may have greater mobility and thus may have narrower, observable lines. The specific structure of the linker between the molecule and the polymer backbone affects mobility and relaxation time and thus can have a large effect on the quality of the spectrum. The best spectra come from supports with long linkers, such as TentaGel, that provide the greatest mobility to the attached molecules. Signals from the more rigid polymer backbone are broad and often not observed.

The best investigated and most generally applicable gel-phase method uses ^{13}C NMR spectroscopy. Proton-decoupled carbon spectra have the highest sensitivity (due to the nuclear

Overhauser effect (NOE)), and are readily interpreted because all peaks are singlets. Carbon spectroscopy suffers from its usual low sensitivity due to the low natural abundance of ^{13}C, making the real-time monitoring of reactions impractical. Gel-phase ^{13}C NMR is therefore most useful when a specific carbon or carbons within the linker or a key building block is labeled with ^{13}C. Changes at or near the labeled atom(s) are very easy to see because of the large ^{13}C chemical shift dispersion. For example, TentaGel resin derivatized with 2-[^{13}C]-glycine can be used to follow imine and thiazoline formation. With ^{13}C enrichment, analysis can be performed on ~20 mg of resin within a 30 min spectral acquisition. Gel-phase NMR also offers the attraction of the use of conventional NMR tubes, solvents, and probes, as contrasted with the magic angle spinning method discussed in the following section, and it is applicable to other nuclei such as ^{19}F and ^{31}P. These nuclei are handy for specific reactions, for example those in which fluoride is the leaving group in a nucleophilic aromatic substitution, but they are not nearly so general as ^{13}C.

Reaction monitoring by gel-phase ^{13}C NMR

High-resolution magic angle spinning NMR

This technique is related to the cross-polarization magic angle spinning (CP MAS) technique conventionally used to observe NMR spectra of solids. The sample is spun at a specific angle relative to the applied field and at a high rate (2-5 kHz). This spinning permits inhomogeneities derived from the different magnetic susceptibilities within a solid sample to be averaged out, narrowing the line widths. However, cross-polarization pulse sequences are not part of high-resolution magic angle spinning (HR MAS) NMR.

^1H NMR spectra can be obtained using HR MAS, an especially attractive feature compared to the gel-phase method. As in gel-phase spectra, molecular mobility as influenced by the linker affects spectral line width. TentaGel-type resins give the best spectra, but decent spectra can be obtained even from molecules bound to Merrifield resin. Line widths for HR MAS ^1H NMR spectra of the *tert*-butyl group of Asp(OtBu)-Fmoc on three resins in three solvents are summarized in Table 3 (facing page). The ^1H NMR spectrum of TentaGel exhibits a large signal around 4 ppm for the ether protons within the polyethyleneglycol chain. Decoupling (effectively, a solvent suppression/pre-saturation pulse sequence) not only removes this signal from the spectrum, it enhances the signals of bead-bound molecules by the nuclear Overhauser effect. Because the NOE is dependent on distance through space, signal strengths of protons in the bound molecule are related to their proximity to the point of attachment to the polyethyleneglycol linker. The NOE makes integration of the spectra not necessarily reliable.

Another attractive feature of HR MAS NMR is that conventional 2-dimensional NMR techniques that are so powerful for structure determination in solution are equally applicable on the solid phase. One simple example is particularly worthy of mention. As shown in Table 3, the

Table 3. Line widths (Hz) of Fmoc-aspartate *tert*-butyl ester in HR MAS ^{1}H NMR

Resin	CD_2Cl_2	DMF-d_6	Benzene-d_6
Wang	10.0	8.7	13.9
NovaSyn-TGA	8.4	10.0	25.1
NovaSyn-TGT	4.1	5.9	19.4

line widths in HR MAS NMR are generally larger than typical J-couplings in ^{1}H NMR spectra, so spin-spin splitting patterns are not readily discerned. However, a 2D J-resolved NMR experiment projects multiplets into the second dimension of the spectrum where they can be easily analyzed. From the perspective of the δ axis, resolved singlets at different chemical shift are seen. From the perspective of the J axis, a triplet and a doublet are seen.

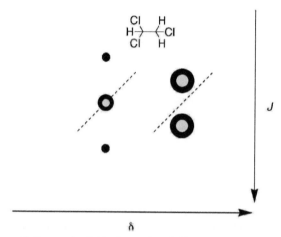

A theoretical 2D J-resolved NMR spectrum

In the practice of HR MAS NMR, a simple one-dimensional ^{1}H spectrum can be obtained with 1-3 mg resin in 16 scans. A one-dimensional ^{13}C spectrum can be obtained in 10 min with 50 mg resin. A specialized, magnetic susceptibility-matched probe accepts the HR MAS rotor, which has a spherical sample compartment, and drives its spinning pneumatically. A NanoProbe with an internal volume of 40 μL, all contained within the RF coils of the probe, can permit spectra to be obtained from single beads in specialized experiments. The HR MAS probe is vastly different from those commonly used to obtain solution-phase spectra. Consequently, HR MAS NMR is a technique most available to laboratories that can dedicate an instrument to on-resin analysis.

On-bead infrared spectroscopy

It is certainly possible to obtain infrared (IR) spectra of polymers. After all, the common IR spectral standard is a polystyrene film. However, obtaining specific IR data on a compound loaded at a fraction of the sites on polystyrene beads is challenging. It is possible to crush support beads and embed them in a KBr pellet, but the quality of the resulting spectra is usually marginal. Preferred is flattening single beads on a NaCl window or a group of beads in a diamond window cell for analysis using IR microscopy. IR microscope attachments are available for FT-IR instruments as a relatively modest investment in hardware.

Because of the complexity of IR spectra, specific functional groups with a unique absorption in a relatively open region in the spectrum are valuable for IR analysis on beads. They can serve as internal standards, or even as tags in molecular encoding. Popular groups that fit this role of tags include triple bonds (nitriles (~2200 cm^{-1}), isonitriles, and acetylenes (~2100 cm^{-1})). Another use of this open window in the spectrum is direct estimation of the yield of a reaction based on the ratio of the IR absorption of the compound undergoing a transformation to a deuterated compound such as a Boc group (($D_3C)_3C)O_2CX$, ~2250 cm^{-1}).

Internal standard for on-resin IR yield analysis

Reactions of specific functional groups, such as cumulenes (isocyanates and isothiocyanates, ~2050 cm^{-1}), that absorb in this region can be easily followed. As in any organic reaction, functional group-specific IR absorptions can be very valuable in following reaction progress. Other absorptions that have been used for on-resin analysis are from NH (3350 cm^{-1}), aldehyde (2750, 1700 cm^{-1}), and OH (3400-3600 cm^{-1}) groups.

Raman spectroscopy/microscopy can be used for similar purposes, and offers the attraction of simpler spectra. It provides a different window into on-bead reactions because it has selection rules opposite from IR spectroscopy. Groups with weak IR bands have strong Raman bands, and vice versa.

Mass spectrometry

While MS can be performed on any molecule cleaved from a resin support bead, the process would be more convenient if the cleavage step were integrated with the analysis step. Photochemically cleavable linkers are the primary means by which this has been performed. A specific linker design can enhance subsequent MS analysis after the cleavage, and this strategy has been termed an analytical construct. An example of such a linker is shown on the facing page. When cleaved from the linker photochemically, the library compound is attached to an amine that is readily protonated. This enhances sensitivity for analysis of the compound in electrospray ionization (ESI) mass spectrometry, which is performed after elution of the construct from the bead. Of course, the mass observed for the synthetic compound must be adjusted to account for the presence of this linker. When the compound is cleaved from the bead with acid in preparation for testing, the analytical construct remains behind on the solid phase. The analysis of compounds using such a linker can be performed in an automated fashion by modern liquid chromatography/mass spectrometry (LCMS) systems. As in IR methods, internal controls are often useful in validating these analytical methods. In this case, the 1,3-diamino-propane linker is a 1:1 mixture of diamines with two ^{14}N-atoms and two ^{15}N-atoms. This results in molecular ions separated by 2 Da. Any ions without such a "doublet" are not derived from the linker. Bromine substitution can be used for a similar purpose because it results in equal intensity isotope peaks for ^{79}Br and ^{81}Br.

An "analytical construct" for MS analysis of compounds on solid support beads

An integrated approach uses the laser of a MALDI-TOF mass spectrometer to directly cleave the molecule from the bead for analysis. Beads are placed on the MALDI sample plate and coated with matrix. The laser is then directed to each matrix-covered bead to obtain a spectrum.

Non-spectroscopic methods

The peptide synthesis literature is a rich source of methods for evaluating reactions on solid phase. There is truly nothing new under the sun. However, the issues in classical peptide synthesis and modern combinatorial chemistry are somewhat different. While single peptides might be prepared on a pilot scale and scaled up to make larger quantities, libraries of compounds are made on smaller and smaller scale to enlarge the diversity of the library.

The simplest of all measures of yield is just the mass gain of a resin upon reaction. Knowledge of the loading and mass of the starting resin and the stoichiometry of the reaction enables prediction of the mass of the resin product, which is a "theoretical yield." Care in washing and removing all of the solvents from the reaction is essential to this method. It is of course also possible to remove the product from the resin to determine its yield. In favorable cases, particularly reactions that involve an uncommon element (anything other than C, H, and O such as S, F, Cl, Br, I), direct combustion analysis of a resin can determine the loading of a product molecule and, by comparison to the starting resin, the yield. Such methods would not be applicable to the analysis of a library where each bead carries a different compound.

The sensitivity of chromatographic methods makes them useful for analysis of compounds prepared at low levels within large libraries. HPLC is the most generally useful technique, but with compounds of sufficient volatility, GC is also applicable, especially when interfaced with MS analysis. A significant difficulty with chromatographic methods is their long analysis time, but technical advances in hardware may address this problem.

With appropriate standards, HPLC can determine product concentration, which can be very helpful in interpreting subsequent assay data. The standardization of the detection method in a HPLC experiment is therefore key. If all members of a combinatorial library contain a common chromophore, for example from a core structure, it is possible to determine product concentration by absorption, but UV is not a general detection method. Two other detection methods more closely approach this ideal. Evaporative light scattering detectors (ELSD) and chemiluminescent nitrogen detectors (CLND) perform well as general molecular detectors. As its name implies, an ELSD measures the light scattered from an aerosol of the column effluent, which is a linear function of the mass of the analyte. The CLND is related linearly to the nitrogen content in a sample. High-temperature oxidation of the column effluent produces nitrogen oxides, which are oxidized by ozone to excited NO_2. Its emission is detected with a photomultiplier to give the detector signal. Naturally, this detector is limited to nitrogen-containing compounds. This is not so great a limitation in pharmaceutical research, as many pharmacologically active compounds include nitrogen atoms. A nitrogen atom might also be included to aid in the analysis of products.

$$\underset{\diagup N \diagdown}{|} \xrightarrow[> 1000\ °C]{O_2} \cdot NO \xrightarrow{O_3} {}^*NO_2 \longrightarrow NO_2 + h\nu$$

Chemistry of a chemiluminescent nitrogen detector

Additional reading

Fitch, W. L. Analytical methods for quality control of combinatorial libraries. *Mol. Diversity* **1999**, *4*, 39-45.

Kassel, D. B. Combinatorial chemistry and mass spectrometry in the 21st century drug discovery laboratory. *Chem. Rev.* **2001**, *101*, 255-267.

Congreve, M. S.; Ley, S. V.; Scicinski, J. J. Analytical construct resins for analysis of solid-phase chemistry. *Chem. Eur. J.* **2002**, *8*, 1769-1776.

Keifer, P. A. Influence of resin structure, tether length, and solvent upon the high-resolution ¹H NMR spectra of solid-phase-synthesis resins. *J. Org. Chem.* **1996**, *61*, 1558-1559.

Problems

1. A large library of molecules based on a pyrrole core related to non-steroidal anti-inflammatory drugs (NSAIDs, like ibuprofen) shown on the following page is desired. Confirmation of the structures of at least a 15 % sample of the library is required before any testing can be done. Mass spectrometry is the obvious method to confirm the structures. One possibility is MALDI TOF, but the library is far too large to use this method for even the sample set. Another possibility is LC/MS (liquid chromatography/mass spectrometry), which has high throughput, but because of their functionality, the test compounds do not ionize well. Suggest a way to get LC/MS evidence of the identity of your products.

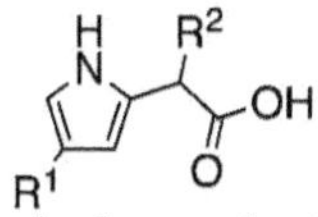

2. Suggest a method to determine whether each of the following reactions has succeeded.

a.

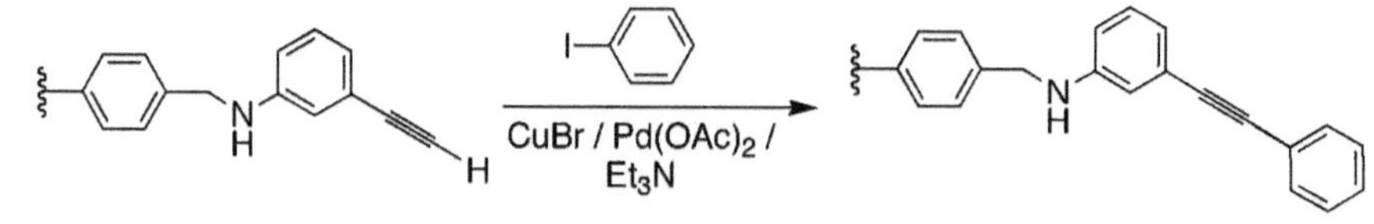

CuBr / Pd(OAc)$_2$ /
Et$_3$N

b.

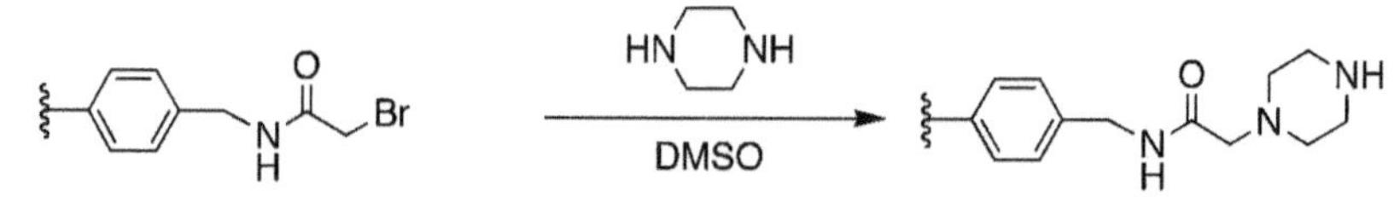

DMSO

c.

t-BuLi

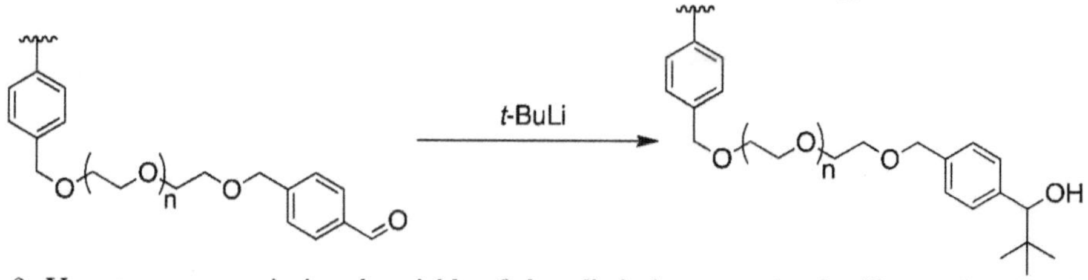

3. You want to maximize the yields of the alkylation steps in the library described in Problem 5.4. Given that ^{13}C-enriched α-bromoacetic acid is commercially available and cheap, suggest a non-destructive method for monitoring these reactions.

4. Which of the following reactions could be followed by gel-phase NMR? by HR MAS NMR?

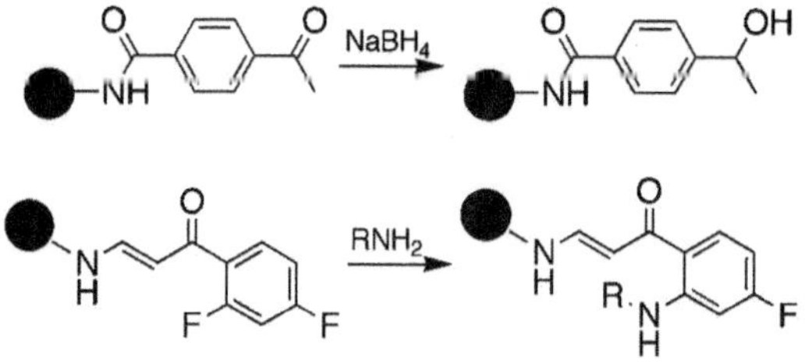

NaBH$_4$

RNH$_2$

Supported Solution-phase Synthesis

Polyethylene glycols

Polyethylene glycols (PEGs) of 3000-5000 Da molecular weight are readily available commercially and well studied. They undergo a solvent-dependent phase transition between tangled and globular forms. They are insoluble in ether, but soluble in many other common solvents. In their soluble form, reactions can be performed in solution even though the molecules are attached to the PEG macromolecule. As in solid-phase synthesis, a high concentration of reagents can be used to make the reaction kinetics most favorable. After reaction, the addition of ether causes precipitation of the PEG-linked product molecule, which can be separated from soluble reagents by filtration.

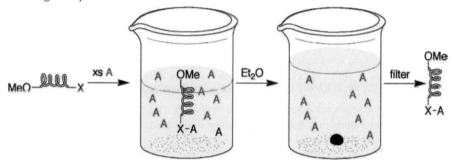

PEG supports combine solution-like reaction conditions with isolation by filtration

PEGs are often obtained with one end etherified and can be prepared with a variety of terminal functional groups, linkers, and handles equivalent to those commonly used in bead-based synthesis. PEG supports have been used in preparation of combinatorial peptide and carbohydrate libraries, among others.

Polyethyleneglycol structure

The virtues of PEG supports for combinatorial synthesis are the solution-like conditions for reaction, which makes translation of a reaction from conventional to the supported version much more straightforward. PEGs offer the ability to analyze reaction products as in other conventional reactions. A drawback of PEG supports is the inability to prepare one-bead, one-compound libraries.

Dendrimers

Closely related to the PEGs, dendrimers constitute soluble macromolecules that can carry a compound during a reaction and permit ultrafiltration or size-exclusion chromatography to separate the product from the small molecules.

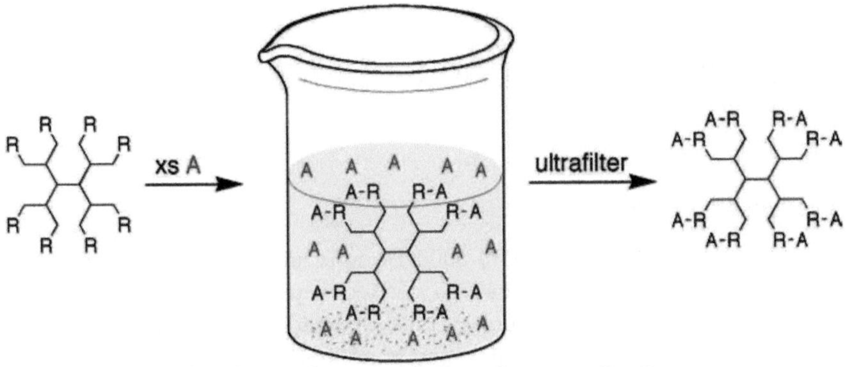

Dendrimer-based solution-phase synthesis

Fluorous synthesis

One of the clichés of chemistry is "like dissolves like." This principle is manifest in the classic solvent extraction, where a polar (water-based) phase and a non-polar (organic solvent) phase are immiscible. The partitioning of reaction products into each phase is a function of their intrinsic polarity and, in some cases, the ionization state of an acidic or basic functional group.

The properties of highly fluorinated molecules seem to fit neither of the foregoing classifications. An example is Teflon, a fluoropolymer that is highly resistant to interactions with both water and organic compounds. Non-polymeric molecules that are highly fluorinated behave similarly. Such compounds may have low solubility in either conventional organic solvents or water, but high solubility in highly fluorinated solvents, which are themselves immiscible with organic solvents or water. Such "fluorous" solvents constitute a third liquid phase that can be used in multi-phase solvent extractions and other phase-partitioning separations.

Conceptually, the linking of a molecule to a highly fluorinated group can play a role in solution-phase chemistry analogous to the role played by the solid phase in supported synthesis. Through the fluorous properties imparted to it by a highly fluorinated group, a molecule can be taken to a different phase from the reagents and by-products in a reaction. The main issues that have concerned chemists are exactly what types of groups can serve as fluorous labels and how they can be used.

A key issue in developing fluorous chemistry was the identification of solvents suitable for highly fluorinated intermediates; two main solutions have been found. The perfluorinated hexanes mixture called FC-72 is commercially available and is a good choice for the fluorous phase in solvent extractions. However, the reagents that might be used in reactions upon fluorous compounds do not commonly come in fluorous form and may not be soluble in FC-72. Therefore, solvents of intermediate "fluoricity" are needed. Good choices for this role are benzotrifluoride (BTF) among the non-protic solvents and trifluoroethanol among the protic solvents. They are miscible with both fluorous and non-fluorous compounds, particularly at elevated temperature.

A simple example serves to demonstrate how fluorous labeling and solvents can be used in synthesis. A highly fluorinated dodecyl bromide is converted to its Grignard reagent in ether and reacts with trichlorosilane. The resulting compound has three times as many fluorines as the starting bromide and is no longer soluble in organic solvents, so FC-72 is used in its conversion to the silyl bromide. This is a reagent that could be used in essentially any of the many transformations of organic synthesis that involve silyl groups, such as protecting groups for alcohols and vinylic/allylic silane reactions.

Preparation of a fluorous benzoic acid

Here, a FC-72/BTF solution of the fluorinated silyl bromide reacts with an aryl lithium generated in ether. Aqueous workup and solvent partitioning between FC-72 and benzene yield the fluorous ortho-thioester in the fluorous phase, which is evaporated. Hydrolysis of the ortho-ester with silver nitrate uses the complex solvent system of BTF, THF, acetone, and water and produces the thioester that is again extracted into a FC-72 phase. Treatment of the thioester with bromine in FC-72 gives the acid bromide, from which the carboxylic acid is obtained by THF/water treatment.

This fluorous benzoic acid constitutes both a novel acylating agent and protecting group (a benzoate equivalent) that when used in place of benzoic acid makes reaction products fluorous. The convenience of phase separation has a price similar to that paid in other solution-phase supported syntheses. That is, a conventional organic molecule requires a large change to make it partition into a different phase (63 fluorines in this label to favor partitioning into fluorinated solvents). The molecular weight of the compound is quite high, which may limit the analytical techniques that can be applied to it, for instance. The high molecular weight also can frustrate the chemist when gram quantities of supported intermediates produce milligram quantities of end-product upon release from the fluorous tag.

This acid has been used in the Ugi four-component condensation, a reaction that will be discussed in greater detail in a later section. Here, the important feature is that by mere extraction of the reaction mixture with FC-72, any by-products from the three non-fluorous components of the reaction are removed. Any unreacted fluorous acid would remain, however, which is why it is used as the limiting reagent. This arrangement has a direct analogy in solid-phase synthesis, where reagents can be used in a high concentration because they partition to a phase that does not contain the desired product.

Fluorous Ugi (Flugi) condensation

In the second step, the aryl-silane bond is cleaved with fluoride ion to release the molecule from the fluorous tag. The FC-72 extraction separates the residual tag from the desired product, which is obtained in high yield and purity.

As interesting and powerful as this methodology promised to be, it is limited by the low solubility of the fluorous-tagged substrate in solvents that dissolve the non-tagged reagents. The number of fluorines in the tag could be reduced to make the tagged compound more soluble in organic solvents, the "light fluorous" strategy. However, large differential solubility of tagged compounds in fluorous vs. organic solvents is the very property that makes fluorous tagging succeed in the solvent extraction stage. Fluorous tags that have fewer fluorines are indeed more soluble, but solvent partitioning into the fluorous phase is incomplete. This dilemma was resolved by replacing the liquid-liquid partitioning with a solid-liquid partitioning. Fluorocarbon reverse-phase chromatographic supports were already known and were exploited for this purpose.

One way these fluorocarbon supports can be used is in HPLC separations. An example of a chromatogram is given on the facing page. A family of amides with identical core structures but a varying length fluorocarbon side chain have quite divergent retention times on this Fluofix 120E column. The retention time is directly related to the number of fluorines in the molecule. None of these compounds, which are so readily separated from one another, effectively partitions into FC-72 in a solvent extraction experiment.

Fluorous reverse-phase silica can also be used in solid-phase extraction (SPE), a process that is similar to but much simpler than chromatography. A mixture of components is added to an adsorbent and eluted with a solvent for which one component has a very high affinity. That component is not retained at all and can be collected directly from the effluent. The other materials that are retained have a very low affinity for the solvent and a very high affinity for the adsorbent, but they might be eluted through a solvent change. It is the significant affinity of fluorinated compounds for this fluorinated solid phase with far fewer fluorines than needed for solvent extraction that enabled this method to be developed.

An example of the use of SPE in fluorous chemistry is shown on the facing page. The amide of a fluorous-tagged proline was prepared by a carbodiimide-based coupling, and the

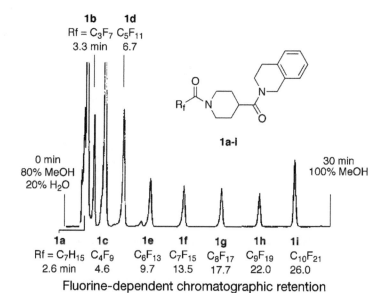

1b 1d
Rf = C_3F_7 C_5F_{11}
3.3 min 6.7
R_f
N
O
N
O
1a-i
0 min
80% MeOH
20% H_2O
30 min
100% MeOH
1a 1c 1e 1f 1g 1h 1i
Rf = C_7H_{15} C_4F_9 C_6F_{13} C_7F_{15} C_8F_{17} C_9F_{19} $C_{10}F_{21}$
2.6 min 4.6 9.7 13.5 17.7 22.0 26.0

Fluorine-dependent chromatographic retention

Reprinted with permission from *J. Am. Chem. Soc.* **1999**, *121*, 9069-9072. Copyright 1999 American Chemical Society.

crude reaction product was placed on top of a plug of fluorous reverse-phase silica gel. The column was eluted first with 80% $MeOH/H_2O$ (to remove the non-tagged products) and then with CH_3CN (to remove the tagged amide). The product was treated with sodium borohydride to reductively remove the fluorous tag, and it was eluted in the first fraction of a second fluorous solid-phase extraction.

CO_2H
N
O
C_9F_{19}
EDC / HOBt
NH
N
O
N
O
C_9F_{19}
89 % yield, 97 % purity
in CH_3CN
$NaBH_4$
N
N
H
O
49 % (as acetamide)
in $MeOH/H_2O$

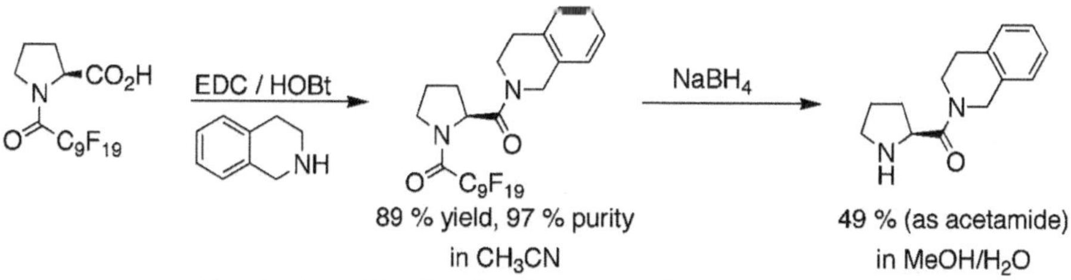

Product purification by fluorous solid-phase extraction

Additional reading

Coe, D. M.; Storer, R. Solution-phase combinatorial chemistry. *Mol. Diversity* **1999**, *4*, 31-38.

Han, H.; Wolfe, M. M.; Brenner, S.; Janda, K. D. Liquid-phase combinatorial synthesis. *Proc. Natl. Acad. Sci. USA.* **1995**, *92*, 6419-23.

Curran, D. P. Fluorous techniques for the synthesis of organic molecules: A unified strategy for reaction and separation, in "Stimulating Concepts in Chemistry," Wiley-VCH; New York (2000), p. 25-37.

Problems

1. Solid-phase peptide synthesis is conducted using perfluorodecanoic anhydride as the capping agent after each coupling cycle. Suggest a method to purify the target peptide sequence from failure sequences, and explain how it works.

2. A U-tube is filled with FC-72. The silyl ether shown below (1 mmol) is placed into an acetonitrile solution in one arm. Fluoride ion in methanol is placed into the other arm. The FC-72 is stirred for 36 h and the methanol solution is extracted with ether. Evaporation of the ether provides pure cinnamyl alcohol (0.9 mmol). Describe what has occurred in this process.

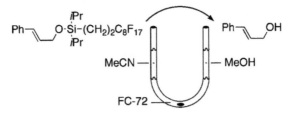

Solution-phase Parallel Synthesis: Solid-phase Reagents, Scavengers, and Extraction

The key principle to be drawn from traditional solid-phase synthesis is that the product and the reagents/by-products of a reaction are found in different phases. The principle applies equally if the reagents/by-products are on the solid phase instead of the product. Thus, an alternative to solid-phase synthesis is solution-phase synthesis using solid-phase reagents. This concept has a number of manifestations, including true solid-phase reagents, post-reaction solid-phase scavenging of excess reagents, and even solid-phase extraction.

Because the molecules in a solution-phase synthesis are not attached to a support, all of the sorting and mixing principles discussed earlier are not applicable. Thus, syntheses using these techniques will produce (primarily) single compounds. They may indeed produce many or all combinations of a set of building blocks, but because no advantage in synthesis efficiency is obtained by conducting simultaneous reactions that produce different compounds, these methods are not so much combinatorial synthesis as they are parallel synthesis.

The ability of parallel synthesis to produce pure, single compounds that can be subjected to all of the conventional criteria for compound identity and purity as has been traditional in chemistry has made it a very popular, perhaps now even the dominant, approach to compound synthesis in the pharmaceutical industry. Value is still placed on making and analyzing many compounds per unit time, so parallel synthesis (also called high-throughput synthesis) is dependent on automation. However, it has required only limited conceptual advances. Likewise, physical methods for analysis of the products of parallel synthesis have become highly automated, as has their biological testing (also called high-throughput screening, abbreviated HTS).

Scavenging resins

A virtue of solid-phase synthesis is that reactions can be performed with an excess of one reactant (so that they are zero-order in the reactant in solution) and go as quickly as possible to completion. The main reason this principle is not widely applied in conventional chemical processes is that removal of the excess reagent can be difficult, or it might lead to side products. A means to remove excess reagents very simply, ideally with a filtration rather than, for example, a solvent extraction, would make this principle of excess reagent applicable to solution-phase synthesis. The concept, then, is that once the limiting reagent has been converted to product, the excess reactants can be sequestered onto a solid phase. This solid phase "scavenges" excess reagent from the solution, hence the name scavenging resins.

A simple prototype reaction serves to better illustrate the concept. A reaction is performed between an amine and an isocyanate to form a urea. The amine is the limiting reagent in this example. After the amine is completely consumed, isocyanate will remain. A polymer resin bearing amines is added. That amine on the solid phase reacts with the isocyanate to form the urea, thereby removing the excess isocyanate from the product solution.

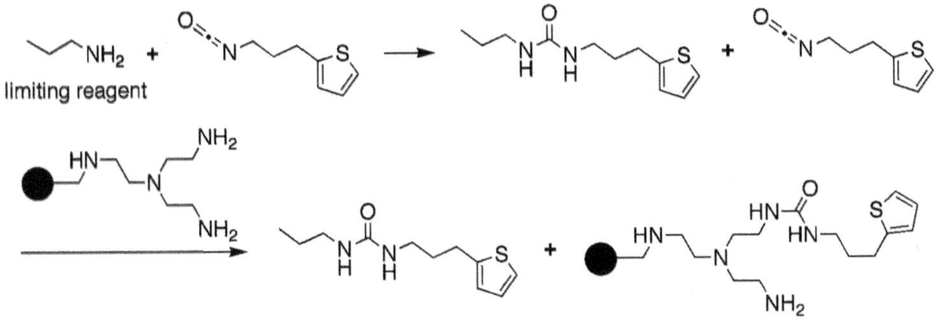

Example of the use of a nucleophilic scavenging resin

This example uses a nucleophilic limiting reagent and a nucleophilic resin. In general, scavenging resins will have the same type of reactivity as the limiting reagent. Ideally, the functional group on the scavenging resin will have similar or greater reactivity than does the limiting reagent. While the experimenter might be willing to wait some time for a reaction with the precious and interesting limiting reagent to go to completion, waiting for completion of the reaction with the polymer resin would not be desirable.

Another example illustrates the use of an electrophilic resin to remove an excess of a nucleophile. The reaction of secondary amines with epoxides gives β-amino alcohols. Epoxides are not very good electrophiles, so the reaction ideally uses the amine in excess. This excess can be removed from the product solution with an isocyanate resin. Note that the product amino alcohol does not react with the resin because the nucleophilicity of an alcohol is much less than an amine.

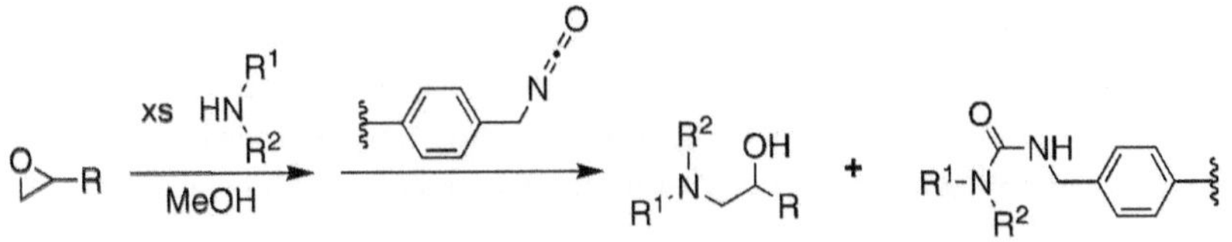

Electrophilic resins remove excess nucleophiles

This example shows the removal of a single reagent with a single resin. If a reaction had multiple unwanted products, multiple resins could be added, each with appropriate reactivity to scavenge each unwanted product.

Ion-exchange resins

The most common method to isolate products and remove by-products in organic chemistry is an aqueous/organic solvent extraction. Often, solvent extraction is used to move reaction products and by-products to different phases based on their acid-base properties. For example, reaction mixtures that include acidic by-products can be readily purified by extraction with basic solutions, which remove the by-products in the aqueous phase as their salts. The execution of solvent extraction is tedious and not readily automated, but resins with the properties of acidic or basic solutions have been known for some time. Because in their ionized forms they can be changed into different ionic forms, they are called ion-exchange resins. Major classes of ion-exchange resins are acidic and basic, with strong or weak sub-classes of each ion-exchange resin.

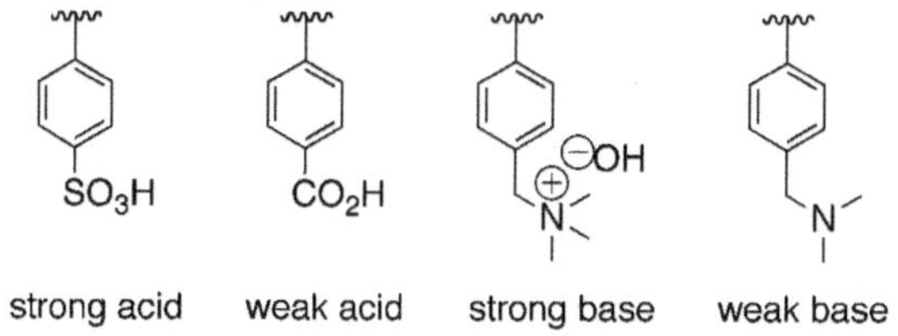

strong acid   weak acid   strong base   weak base

Classes of ion-exchange resins

Carboxylic acid resins are weak acids. Sulfonic acid resins (obtained by sulfonating polystyrene) are strong acids. Amine resins are bases, with quaternary ammonium hydroxides being strong, and simple tertiary amine resins being weak.

A simple example shows how such resins can be used in purification. The reaction of an acid chloride with an excess of amine produces the amide. On loading the reaction mixture onto a strong cation-exchange resin, the amine is protonated by the sulfonic acid. Charge-charge interactions prevent the ammonium ion from being eluted from the resin, while the amide reaction product, which is neutral, is not at all retained.

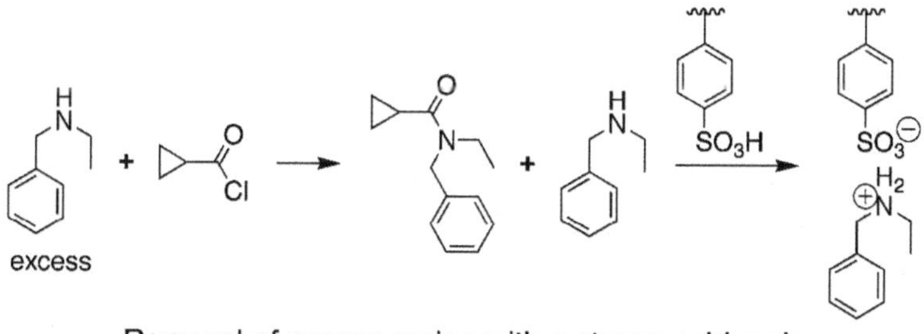

excess

Removal of excess amine with a strong acid resin

It is also possible to use ion-exchange resins to sequester by-products that are not themselves acidic or basic by judicious choice of "adapters," quenching reagents that are basic or acidic. In the example given on the following page, the excess isocyanate reacts with a diamine as a quenching agent. The resulting urea product still has a basic nitrogen so that it can be removed by a strong acid resin. The desired urea is not affected by the acid.

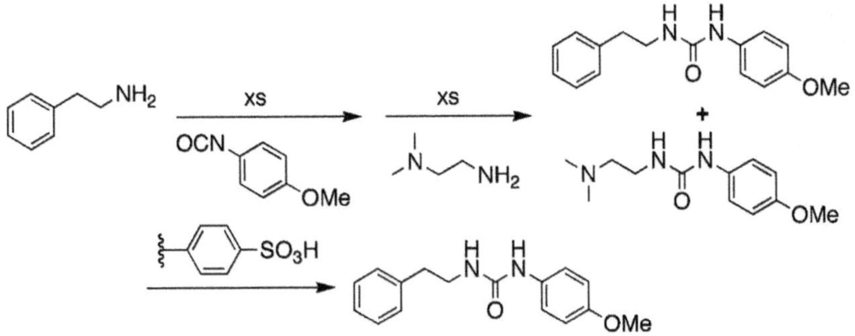

Basic quench reagents can permit by-product removal by acidic resin

An even more straightforward use of ion-exchange resins is as replacements for acidic or basic aqueous quenching solutions. In the Grignard reaction shown below, the carboxylic acid resin substitutes for the normal acidic workup. It also binds and removes magnesium ions. The primary amine resin is added to sequester any unreacted aldehyde as the imine. This resin substitutes for an equivalent process in solvent extraction: formation of a water-soluble bisulfite addition adduct from an aldehyde.

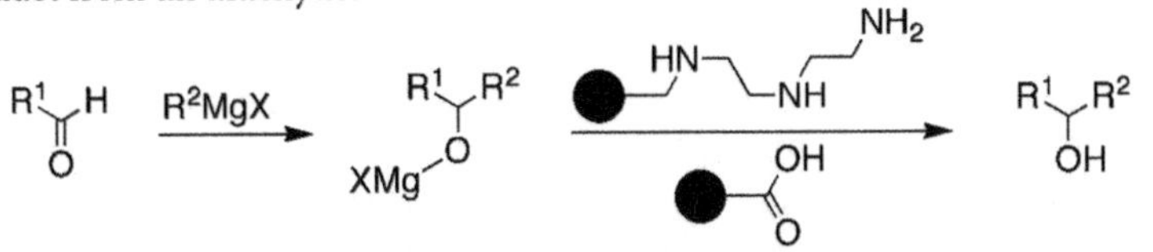

Resin-based quench of a Grignard reaction

All of these methods have in common the substitution of a filtration step for a conventional solvent extraction step.

Supported reagents

An alternative approach to maintaining the phase separation of reagents and products is to have the reagents on the solid phase rather than the reactants. Provided the reaction goes to completion, removal of excess reagents is merely a filtration, as in conventional solid-phase synthesis, but the desired product is in the filtrate, not on the support. This feature is desirable due to the already discussed issue that the reactivity of compounds may be adversely affected by their immobilization on the support. Of course, the reactivity of a reagent may also be diminished by immobilization.

The value of solid reagents has long been recognized in organic synthesis. Many examples of heterogeneous reactions are known, particularly based on heterogeneous metallic catalysts such as palladium (and other noble metals) on carbon used in hydrogenation reactions. A well-recognized issue in heterogeneous reactions is the phase separation that may create slow mass transport between the bulk solution and the sites on the solid phase at which reaction may occur. A number of methods have been suggested to enhance mass transport, such as ultrasonication. Many advances have been made in the reagents commonly used in organic chemistry by their conversion to supported reagents. Polymer supports can be used, but other solid phases that are also commonly used as supports include carbon, silica gel, alumina, clays, and diatomaceous earth (Celite®). A few examples will be discussed of solid-phase reagents that represent a small sample of the large number of such species.

A polymeric version of pyridinium perbromide can be made from the commercially available polyvinylpyridine (**PVP**). Like pyridinium perbromide, it is a useful reagent for electrophilic bromination α to ketones. It even shows the same discharge of bromine color as the reaction proceeds.

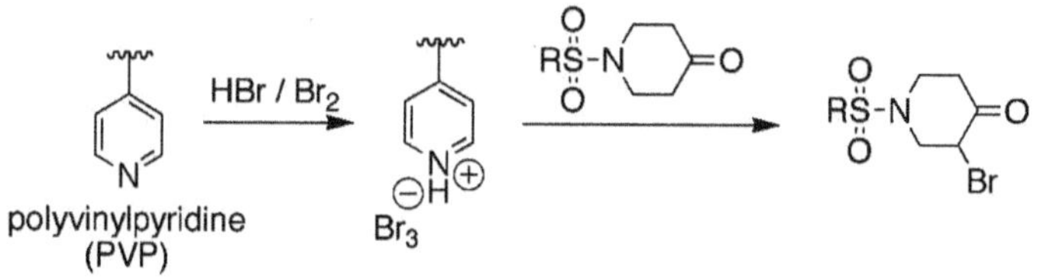

polyvinylpyridine
(PVP)

Synthesis and use of polyvinylpyridinium perbromide

Ion-exchange resins can be used as supports for charged reagents that would commonly be used in aqueous solution. For example, borohydride can be exchanged onto the strongly basic resin shown below. The reductive amination of tyramine and veratraldehyde performed in methanol gives the secondary amine in 90% yield.

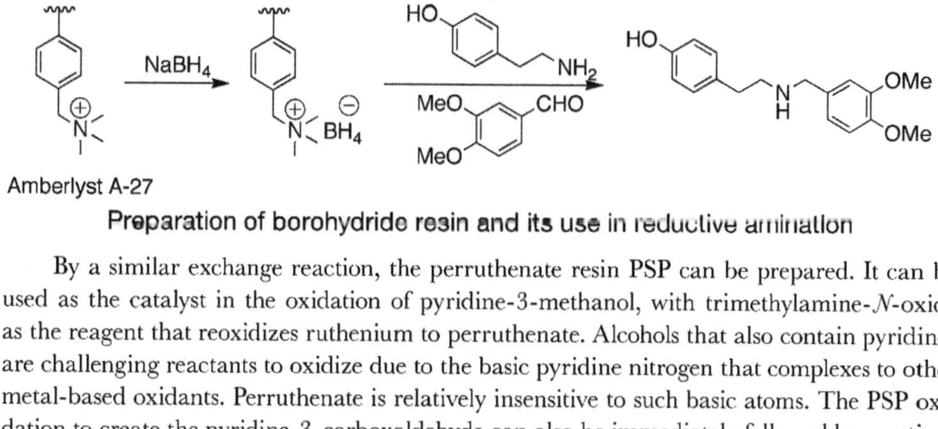

Amberlyst A-27

Preparation of borohydride resin and its use in reductive amination

By a similar exchange reaction, the perruthenate resin PSP can be prepared. It can be used as the catalyst in the oxidation of pyridine-3-methanol, with trimethylamine-*N*-oxide as the reagent that reoxidizes ruthenium to perruthenate. Alcohols that also contain pyridines are challenging reactants to oxidize due to the basic pyridine nitrogen that complexes to other metal-based oxidants. Perruthenate is relatively insensitive to such basic atoms. The PSP oxidation to create the pyridine-3-carboxaldehyde can also be immediately followed by reactions using other solid-supported reagents.

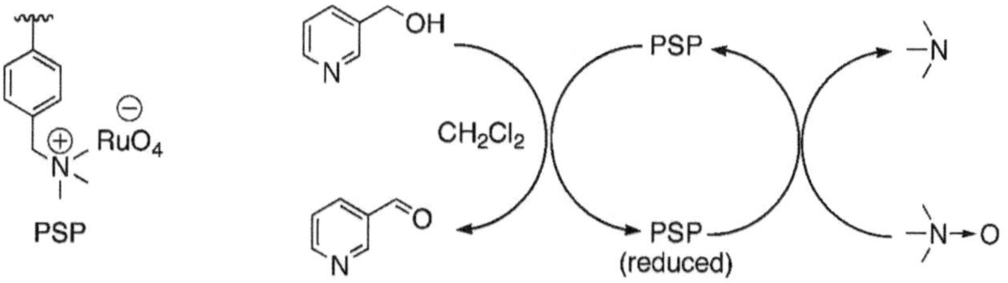

PSP

Use of polymer-supported perruthenate (PSP) in alcohol oxidation

An example of a reagent supported on a porous solid is silver carbonate on Celite®, also known as Fetizon's reagent. It can be used as an oxidizing agent for a hydroquinone in a parallel high-throughput synthesis.

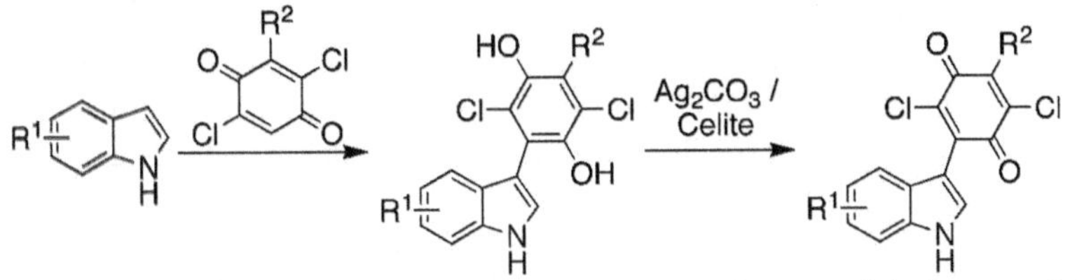

Preparation of (3-indolyl)benzoquinone libraries using a solid-phase oxidant

Fluorous reagents

Like solid-phase supported reagents, fluorous-phase supported reagents permit ready separation of products from by-products. Fluorous techniques have been applied to organic reactions that are well-known for their difficulty in removing reagent residues, such as the tri-*n*-butyltin hydride reduction of alkyl halides. A further enhancement in the reaction given below was using the tin hydride in catalytic amounts and adding a reductant, sodium cyanoborohydride, to convert the stannyl bromide product back to the tin hydride. Making the tin hydride fluorous permits its ready removal using the solvent extraction methods discussed previously. In this case, because of the inorganic component of the reaction, an aqueous phase is needed to accept the salts, etc., formed in the reaction. Thus, a three-phase solvent extraction is performed.

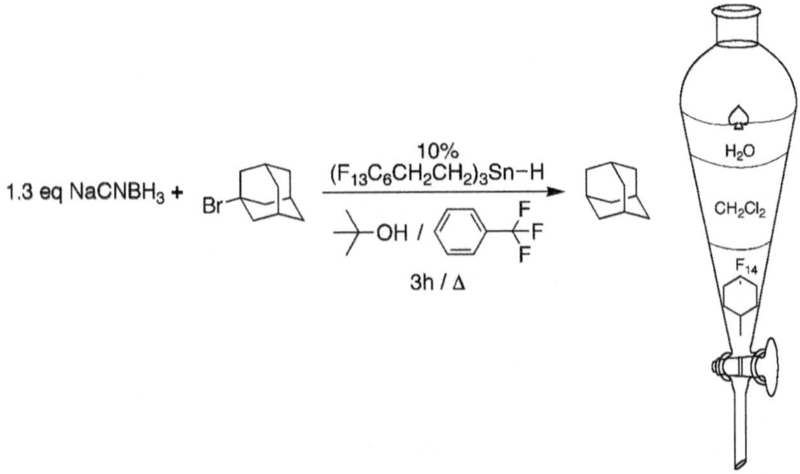

Fluorous tin hydride solvent extraction

Solid-phase extraction

Another technique to increase the throughput in a solution-phase synthesis is to eliminate laborious aqueous solvent extraction routines designed to purify reaction products. Solid-phase extraction was earlier discussed in the context of separation of fluorous and non-fluorous reaction products by adsorption onto fluorous reverse-phase silica gel. A very simple and practical method to eliminate solvent extractions merely adsorbs a limited quantity of an aqueous solution onto diatomaceous earth (Celite®) and passage of an organic solution of the reaction mixture through it.

Gas-phase separation

Evaporation is a final method of purification that is almost too obvious to mention because it is used so pervasively. However, in favorable cases, reagents that give only volatile by-products can be very helpful in directly providing a pure product. A simple example is the Mitsunobu reaction with di-*tert*-butyl azodicarboxylate (DTAB). Unreacted DTAB as well as the hydrazine product of the desired reaction are converted to isobutylene, carbon dioxide, and either diimide or hydrazine by removal of the Boc group with TFA.

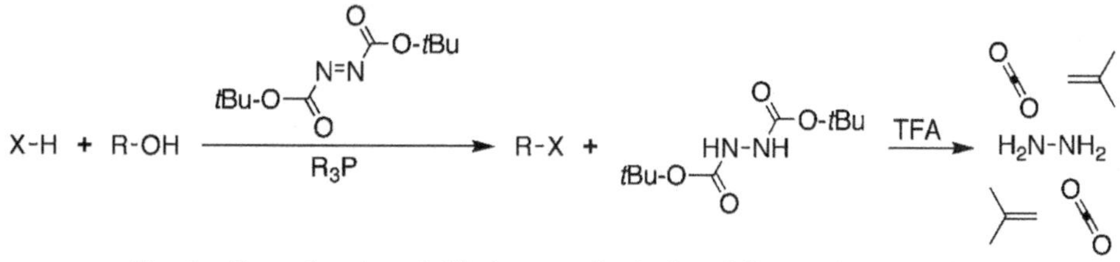

Production of only volatile by-products in a Mitsunobu reaction

Additional reading

"Supported Reagents: Preparation, Analysis, and Applications." Clark, J. H.; Kybett, A. P.; Macquarrie D. J., Wiley, NY (1992).

Ley, S. V.; Baxendale, I. R.; Bream, R. N.; Jackson, P. S.; Leach, A. G.; Longbottom, D. A.; Nesi, M.; Scott, J. S.; Storer, R. I.; Taylor, S. J. Multi-step organic synthesis using solid-supported reagents and scavengers: a new paradigm in chemical library generation. *J. Chem. Soc., Perkin 1* **2000**, 3815-4195.

Problems

1. Given assistance by the amine shown, describe a method by which a solution-phase parallel synthesis of phenylthioureas could be performed.

$$H_2N(CH_2)_2C_{16}F_{33}$$

2. Describe the preparation and use of an ion-exchange resin to perform the following transformation.

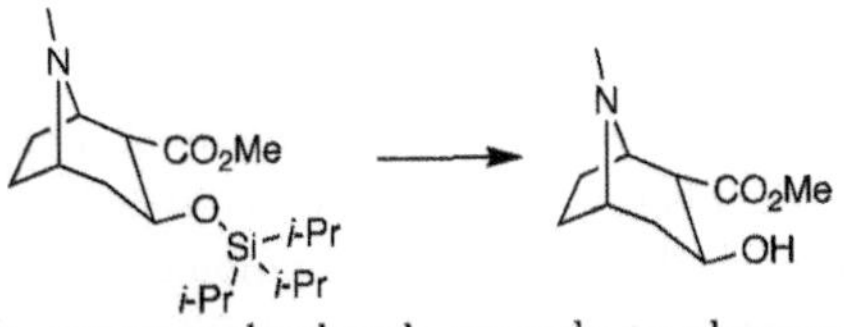

3. Polycyclic aromatic compounds absorb strongly to charcoal. Using this information, describe how the protecting group shown could be used in solution-phase synthesis of the dipeptide aspartame.

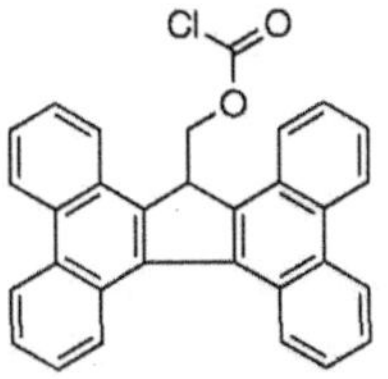

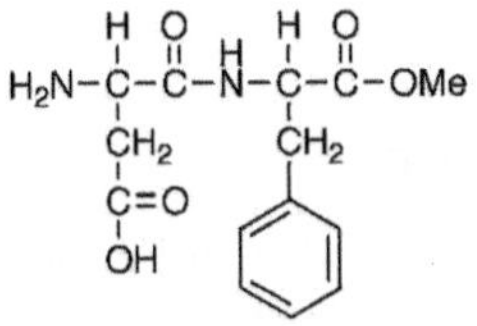

aspartame

4. Suggest a phase separation scheme for the products of the following reaction, identifying the phase to which each compound will go.

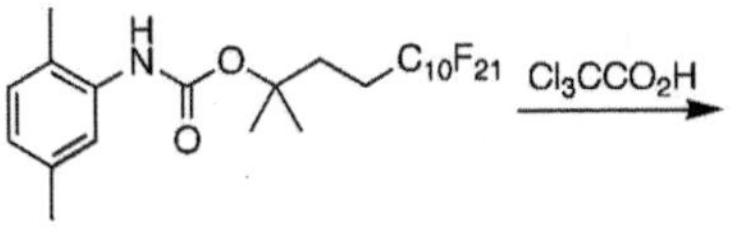

Multi-component Reactions

Most reactions involve two stoichiometric reactants and give one product. Reactions in which more than two reactants come together to produce a product are called multi-component reactions. Because the likelihood of a true termolecular (or higher order) reaction is very low, multi-component reactions generally involve two reactants forming a reactive intermediate that then reacts with another reagent to form products.

$$A \; + \; B \longrightarrow P \quad \textit{versus} \quad A \; + \; B \longrightarrow I \xrightarrow{\; C \;} P$$

In some ways, it is natural to apply multi-component reactions to combinatorial chemistry. In a usual two-component reaction, the number of possible combinations increases with the square of the number of building blocks used for each reactant, whereas in a three-component reaction it increases with the cube of the number of building blocks. This relationship is analogous to that between the number of combinations of dipeptides versus tripeptides. However, it will be recalled that the split/couple/mix method relies on *sequential* addition of building blocks. It is not possible to use split/couple/mix methods when more than one type of chemical diversity element or building block is being introduced in the same reaction step. Multi-component reactions are most applied in solution-based parallel synthesis. They can also be applied in pool synthesis where the positional scanning or indexing methods are used.

The classic multi-component reaction, the Ugi four-component condensation, has certainly been well recognized and utilized for its ability to unite building blocks from four diverse and yet readily available functional group families: carboxylic acids, carbonyls, amines, and the much rarer isonitriles. The products of an Ugi reaction are N-acyl α-amino amides, and thus are closely related to peptides. This is a virtue both because of the important biological properties of peptides and because the differences between Ugi products and native peptides may make the former more valuable as drug candidates. Ugi condensation products might even be considered an alternative oligomer set.

Ugi four-component condensation

The Ugi reaction has an interesting mechanism. The formation of an imine between the carbonyl (such as an aldehyde, as shown here) and an amine is a classical reaction in organic chemistry. The initial formation of an imine in the reaction means that preformed imines or imine-like functional groups may also be used in Ugi-type coupling processes. The imine is protonated by the carboxylic acid, making it electrophilic as the imminium ion shown. The isonitrile group is dipolar and carries significant negative charge character on its carbon. Therefore, this carbon can act as a nucleophile toward the imminium ion. The product of this process is an iminacylium ion that can react with the carboxylate ion to form what is effectively a mixed anhydride between a carboxylic acid and an imide. This is a reactive acylating agent, and the nitrogen that was made nucleophilic by the addition of the isonitrile to the imminium ion can perform an intramolecular acyl substitution reaction to give the product. That so much can happen in a single reaction, and often quite efficiently, is amazing.

Mechanism of the Ugi condensation

As earlier discussed, classical synthesis methods are used to prepare peptides. In forming polypeptides, only N-C bonds are made, and they are made sequentially. In the Ugi reaction, a C-C bond is made in the same reaction as an N-C bond. The formation of C-C bonds is the greatest challenge of organic synthesis, and any method that can be reliably used to form C-C bonds in combinatorial chemistry will be a valuable one.

Peptide synthesis versus Ugi condensation

The application of the Ugi reaction to combinatorial chemistry has been broad. Several examples of the types of creative additions to the Ugi arsenal are discussed following.

The Ugi reaction can be used in solid-phase synthesis by making one of the four reactants resin-bound. As resins are not as available as the four classes of compounds in an Ugi reaction, making one of the four resin-bound limits the diversity of the library to some extent. It does allow the product to be purified by filtration as in conventional solid-phase synthesis. Since the product is assembled on the resin from reactants in solution, this process is sometimes called resin capture.

An example of the Ugi reaction with *resin capture* is given on the facing page. The amine component is on the solid phase, in this case as Rink amide resin. Upon cleavage of the product from the support, Rink amide resin creates a simple amide, so that will be the amine functionality present within each product. The other reactants (acid, isonitrile, carbonyl) can be of any structure, and were used in a 10-fold excess in this example. After removal of these reagents, the product is cleaved from support with TFA.

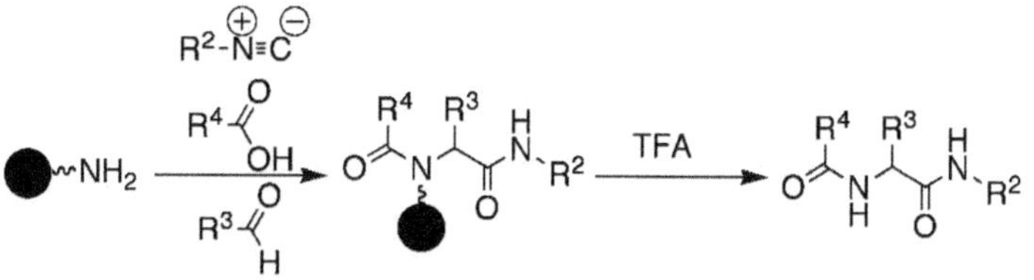

Ugi reaction with resin capture

In this particular process eight aldehydes and twelve acids were used to constitute one 96-well plate.

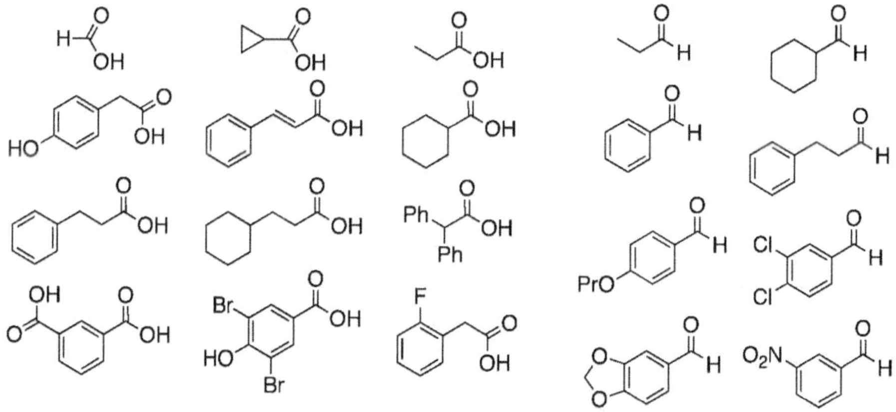

Diverse aldehydes and acids for an Ugi plate

Only one isonitrile, ethyl isocyanoacetate, was used in this plate. A representative structure generated from this library synthesis is given below. While this process generates 96 different compounds, examination of their structures shows that many of the building blocks (and resulting Ugi products) are quite similar. This issue draws attention to the problem of library design that will be covered in the next section.

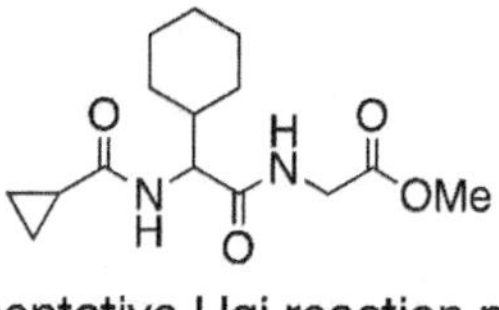

Representative Ugi reaction product

Because the aldehyde and amine in an Ugi reaction form an imine in situ, it is possible to substitute an imine for these two components and accomplish the preparation of an acyl aminoamide. This option provides access to compounds that are not available by reactions of an amine and an aldehyde, such as heterocycles. In the example on the following page, the amino-aldehyde corresponding to the dimethylthiazoline starting material would not be stable because it is an N,S-hemiacetal. The reaction on the following page is also more complex than the standard Ugi reaction in that the initial condensation product undergoes further reaction.

The nucleophilic sulfur of the thioamide adds intramolecularly to the reactive vinyl sulfone moiety to form an intermediate thiazoline. It can aromatize by loss of the sulfinic acid, which is a good leaving group.

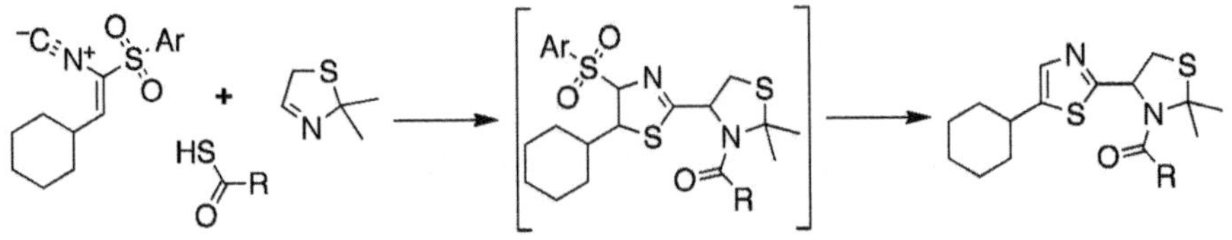

An imine Ugi reaction

Another multi-component reaction that has been used in combinatorial chemistry is the Passerini reaction. It is very similar to the Ugi reaction except the amine is omitted so that α-acyloxyamides are produced.

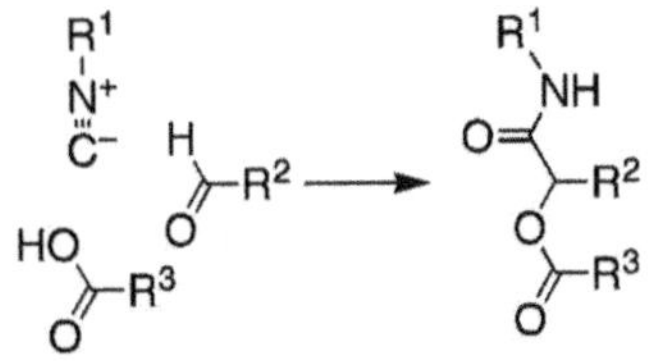

Passerini reaction

The Passerini reaction was used to assemble one of the basic units of the natural product azinomycin, shown in the box on its structure.

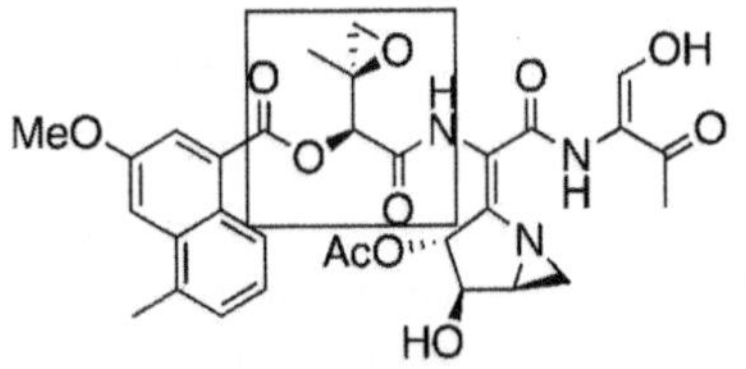

Azinomycin B

Resin capture of the products was also used in this study. They were released from the support photochemically. The success of the different combinations of aldehydes and isonitriles is indicated in the matrix by reaction failure (x), adequate reaction (~), and good reaction (√).

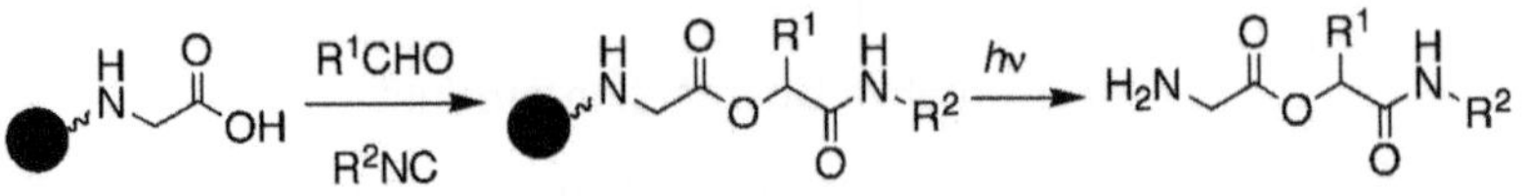

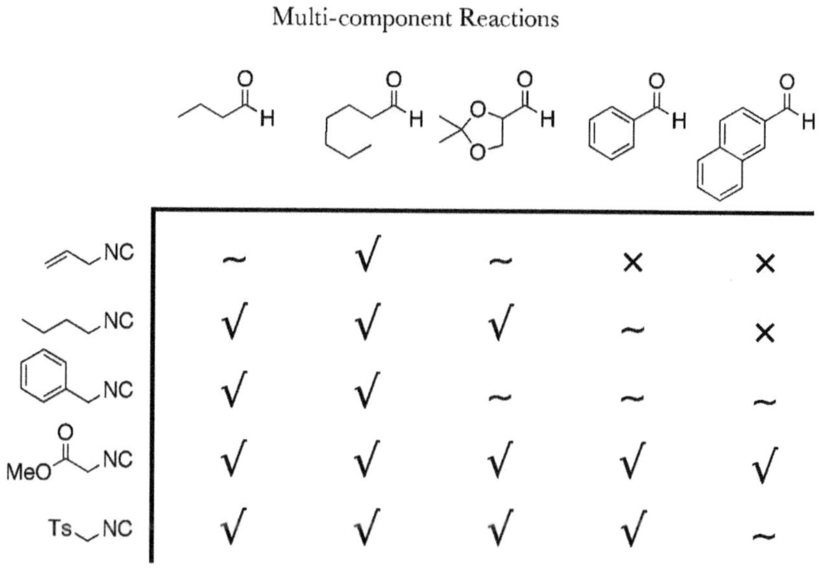

Library of Passerini reaction products

The multi-component reactions of isonitriles given below are so spectacular, they provide a challenge even to identify the sources of particular atoms of the product, much less to provide a cogent reaction mechanism. To this end, an intermediate is given.

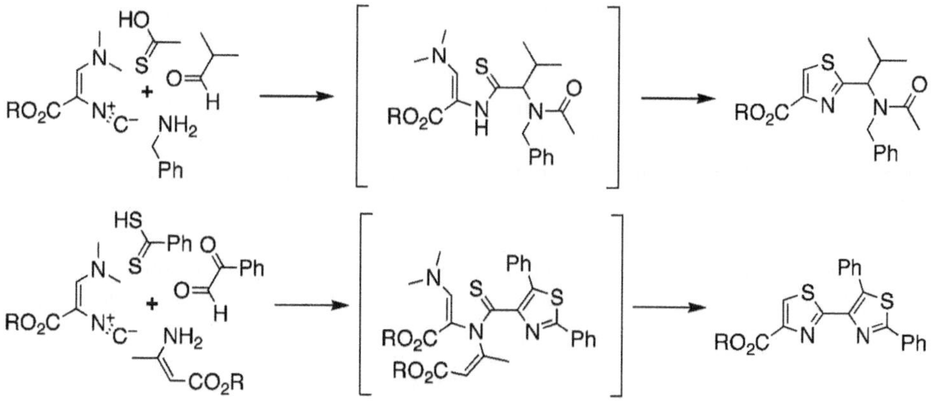

Complex multi-component reactions

Another reaction involving isonitriles can be used to make polyheterocyclic compounds that have a very drug-like appearance. Two diversity elements are shown in this reaction. Varied 2-aminopyridines would be necessary to make the library as diverse as possible.

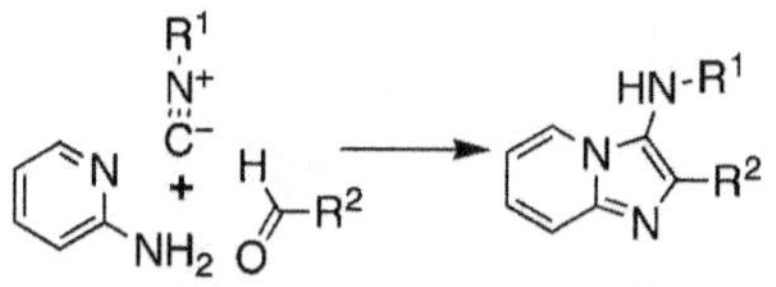

Multi-component assembly of benzoimidazoles

Another multi-component reaction that has been applied to combinatorial chemistry is the Biginelli condensation, which uses ureas, β-ketoesters, and aldehydes to form pyrimidines. This reaction has been performed with a fluorous tag attached to the urea to facilitate purification of the condensation product. The fluorous silyl group is removed with fluoride ion to give the benzoate.

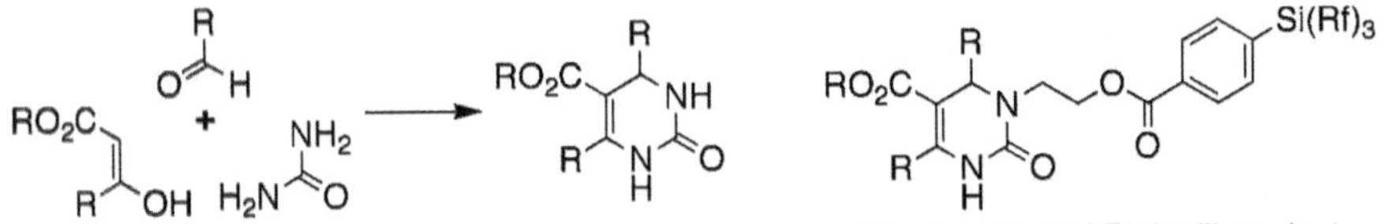

fluorous tagged Biginelli product

Biginelli condensation

The formal cycloaddition between alkenes and aldimines (formed from anilines and aldehydes) can lead to interesting heterocycles.

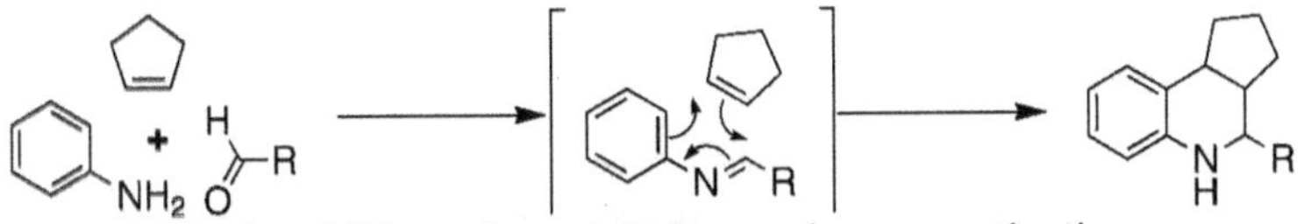

Cycloaddition of an aldimine and rearomatization

Many other well-known reactions in organic synthesis, such as the Mannich reaction, unite multiple components and can be used in combinatorial synthesis. Each component may not bring diversity to the product, but the reactions are still classified as multi-component condensations. A good example is the Hantzsch synthesis of dihydropyridines, an important route to calcium channel blocking agents (such as nifedipine). It involves the acid-catalyzed condensation of aldehydes with two equivalents of a β-dicarbonyl compound and ammonia. The Hantzsch products have two independent points of diversity. These reactions can be applied to solid-phase synthesis by using a cleavable amine resin as an ammonia equivalent. A β-diketone condenses with the amine to form a β-aminoketone. This intermediate reacts with an aldehyde and a β-ketoester to form the dihydropyridine on the solid phase. Oxidation of this species removes the compound from the support and gives the fully aromatic pyridines. Performing this

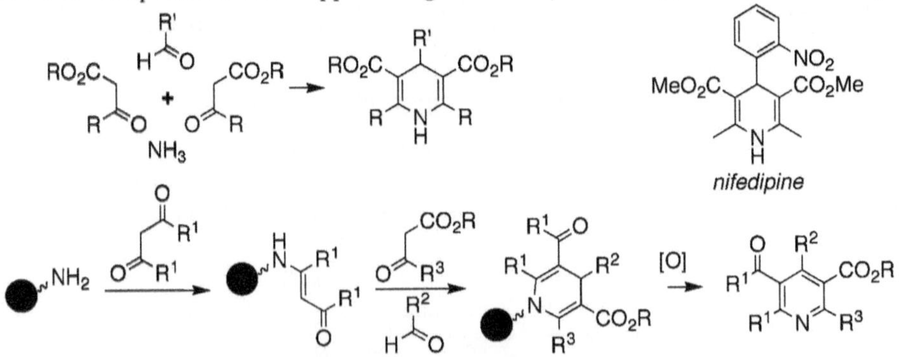

nifedipine

Hantzsch route to dihydropyridines and highly substituted pyridines

reaction on the solid phase permits the incorporation of two different β-dicarbonyl compounds, which is not practical in solution.

Prize Winning? A paradigm of combinatorial chemistry is a template molecule to which peripheral diversity elements are added. As was discussed earlier, this tradition seemingly is required so that a uniform linking chemistry will succeed with all building blocks. As a result, some parts of the molecule are not diverse at all, in fact they are identical. This problem was first recognized by scientists at the MorphoChem company, which is one of the leading prac- titioners of multi-component reactions and an advocate for "high diversity" combichem. At a 2001 scientific meeting, they posed a challenge to combinatorial chemists to create a 3 × 3 × 3 library where the peripheral groups are held constant and the backbone of the library var- ies, rather than the reverse. The person to accomplish this will get the MorphoChem Prize.

Additional reading

Gordeev, M. F.; Patel, D. V.; England, B. P.; Jonnalagadda, S.; Combs, J. D.; Gordon, E. M. Combinatorial synthesis and screening of a chemical library of 1,4-dihydropyridine cal- cium channel blockers. *Bioorg. Med. Chem.* **1998**, *6*, 883-9.

Problems

1. Provide the reaction pathway for the following process.

2. Suggest a one-step synthesis of xylocaine, the topical anesthetic.

3. Suggest a mechanism for this multi-component reaction:

4. Propose a reaction pathway for the following transformation, including the identity of intermediate A.

Chemical Informatics, Diversity, and Library Design

Some of the specific libraries that were prepared in the preceding sections raised the question "what constitutes diversity?". It is always possible to compute the number of compounds in a synthetic library, but perhaps this does not really reflect *chemical* diversity. It will be rare that such a group of compounds is truly diverse in *structure*. One of the tradeoffs that combinatorial chemistry makes to create large numbers of compounds is that they all have similar functionality, at least the functionality involving bond formation. This is necessary so that reaction conditions are compatible among different building blocks. Peptides all use the amide bond to link together different building blocks, for example. For the parts of the molecules that are different, it is necessary that they be as different as possible in order to make the library as diverse as possible.

Given this concern, considerable thought has gone into how to create diverse libraries, and even to measuring diversity. A key concept has been introduced to capture this idea, *chemical space*. This concept refers to the fact that molecules have a wide variety of structures and properties, each composing a different value in one dimension of an *n*-dimensional space. People can readily conceive of spaces up to three dimensions, but dealing with higher dimensions is only tractable mathematically. The properties of a molecule that could be relevant in combinatorial chemistry are truly limitless, but some that might be used in considerations of diversity include molecular formula, molecular connection table, functional groups, molar volume, octanol/water partition coefficient (log P), topology, and number of rotatable bonds. The likelihood of a compound having a desired activity based on its particular molecular properties might also be included (i.e., for orally active drugs, molecular weight < 700 Da).

Molecules are likely spread rather sparsely through chemical space, like stars in outer space. The structures and properties of molecules are certainly not continuous functions. For example, bond lengths and angles adopt discrete values based on bond order and hybridization, so some geometric arrangements of atoms will be inaccessible in *any* type of molecule. It is thought that related molecules will be found in clusters within chemical space, but that different types of molecules will be found in different parts of chemical space. Molecules are still unique, even though there are many of them. The daunting task we face is to make and test the "right" one for our task.

> *How many molecules are possible?* Estimates based on bonding rules for the elements common in organic compounds, reasonable molecular size for pharmacological agents, etc., suggest that more molecules are possible than the number of atoms in the Universe. Obviously, *all* cannot be made and tested, so design/sampling strategies are important.

Strategies

If the type of molecule needed for a task is completely unknown, chemical space might be sampled sparsely in the library design. If some particular property or structure is required, it can be incorporated into the design. Practically, a combinatorial library of "all" compounds cannot be prepared, even if they are limited by some criterion. If a lead structure is known when the library is being designed, diversity around that lead can be generated to form a focused library.

initial screen

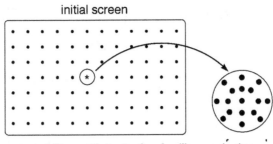

Sparse and focused strategies for library design

Libraries must always be organized around a specific chemical structure (*template, scaffold,* or *core*) that facilitates a reliable synthesis. TAT is a structural template that enables libraries to be readily prepared from trichlorotriazine, for example.

Triaminotriazine (TAT) template

It is also important to keep in mind that, at least for industrial research, the building blocks used in a combinatorial synthesis must be available by purchase (almost exclusively). While perhaps a few key templates might be prepared by a multi-step chemical synthesis, if the compounds that are incorporated into the library for diversity each required preparation, the time spent on preparing building blocks would far exceed the time spent preparing the library. It is important to have complete access to the catalogs of all chemical suppliers, because often some of the more interesting and less common compounds that offer more diverse structures will not come from the major suppliers.

Several resources are available to find commercially available compounds that fit into the chemistry being used in the creation of a library. Basically, this means the ability to search commercial compounds by chemical sub-structure. The best known resource for this task is the Available Chemicals Directory, or ACD. It is primarily available to large industrial subscribers. The SciFinder Scholar program made available to universities by the American Chemical Society also permits searching for commercially available compounds by criteria that include sub-structure. At the laboratory level, CambridgeSoft offers the ChemACX program, which uses a relational database application to search a CD of commercial compounds (updated yearly). There is also a web-based version of ChemACX that uses the ChemDraw Plug-in and searches the current ChemACX database. If compounds located on the web are also on the most recent CD, they can be seen but the prices and suppliers cannot be. If they are newer than the CD, all information is available from the web site. The URL for ChemACX is:

http://chemacx.cambridgesoft.com/chemacx/index.asp

The accessibility of commercially available compounds by sub-structure is a significant enhancement to the capabilities of any laboratory, even those that are not performing combinatorial chemistry.

Representative flowchart for a library design

• Identify functional group classes needed for the chemistry (the TAT template uses primary amines);

• Find, using computer databases, all of the commercially available compounds of that class (a ChemACX search in 2002 gave 1334 primary benzylic amines, which likely includes some repeats);

One page from the results of this search is shown below.

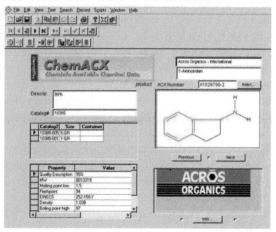

Results screen from ChemACX search for primary benzylic amines

• Group/filter the hits based on traits like chemical reactivity (dependent on the chemistry) that are crucial to the preparation of the library (e.g., 2,4-dinitroaniline is a very weak nucleophile, and would not be a good reactant in the nucleophilic substitution chemistry used with the TAT template). One successful approach has involved selecting building blocks that are maximally dissimilar to one another. Other filters, like price, are applied;

• Create using the computer a *virtual library*, which includes all possible compounds that can be formed from the appropriate building blocks (assuming adequate reactivity at each point of diversity on TAT, that would be 1334^3 (2.3 billion!) compounds);

• Calculate molecular properties for each compound in the virtual library, and place them into n-dimensional space;

• Use principle components analysis (PCA) or other related statistical techniques to project the n-dimensional space involving shape, surface traits (charge, H-bonding, and hydrophobic character), and other properties into a two- or three-dimensional space that can be visualized and analyzed. An example is provided from a library design software package.

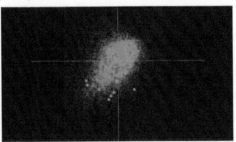

Molecular clustering in 3-dimensional space by the ChemSpaceShuttle program

Principal components analysis is able to determine which parameters in the raw data affect the measured response (output) of the system in what proportion and can readily account for the common occurrence where one property varies in parallel with another property.

• Constitute the library from the remaining compounds, including definition of the size of the library that actually can be made and tested.

For finding an initial lead, it might be desirable that chemical space be covered sparsely, so compounds that are similar would be eliminated. In other words, two different compounds might have very similar calculated properties. It would be a waste of effort to prepare both if they were not expected to give different outcomes in the library testing phase. For situations in which a lead structure is already identified, library design might involve choosing molecules from the virtual library that are quite similar to the lead. If the initial screen finds a lead in the sparse chemical/property space for the specific template, the second screen might focus around that hit, looking for similarity. While the initial screen might focus on binary answers (either a compound or pool is active or it is not, a determination that may require only a single assay), the screening of secondary, more focused libraries might involve determination of an EC_{50} (the concentration effective at providing 50% of the maximum desired effect), which requires assays at several different concentrations.

Library design influences pooling strategies. A very diverse library would be expected to provide few hits. One theory is that grouping similar molecules in the same pool would result in a higher probability to find active pools than if similar hits are distributed among a number of pools. Another theory is that maximizing diversity in a pool improves the chance that activity comes from a highly active unique compound, rather than lots of mediocre compounds.

Several approaches have been applied to measuring diversity. Some useful metrics that have been found include CoMFA (comparative molecular field analysis) and 2D fingerprints of the diversity groups that are not part of the template. Other factors that are used in quantitative structure-activity relationship (QSAR) studies (melting point, molecular weight, dipole moment, refractive index) might also be incorporated into library design. If a 3-dimensional structure of the target protein were available, the members of the virtual library could be tested computationally for possible interactions with the target to select the most likely hit compounds to synthesize.

The presentation of multiple properties of molecules within a library (virtual or real) for ready perception is challenging. One solution is the "flower plot" developed at the Chiron company. In the example provided below, there is one petal for each of the five chemical functionality descriptors, five shape descriptors, five receptor recognition descriptors and the computed log P.

Flower plots of combinatorial building blocks

Lipinski's Rules

Another criterion that can be applied to a library intended for drug discovery was developed by Lipinski of Pfizer. He provided a set of empirical "rules" that reflect the types of chemical structures/properties that are most often found in orally active pharmaceuticals. The rules are related to the solubility of molecules and their ability to penetrate through biological membranes, which apparently can limit the activity even of compounds that are potent *in vitro*, where these issues are irrelevant.

Lipinski's rules state that poor absorption or membrane permeability is more likely when a compound has:
- > 5 H-bond donors (sum of OHs and NHs);
- molecular weight > 500 Da;
- logP > 5;
- > 10 H-bond acceptors (sum of Ns and Os).

An additional criterion for valuable pharmaceutical compounds not related to permeability or solubility is that they should not have > 5 rotatable bonds. This criterion is likely related to the fact that molecules are not conformationally flexible when bound to their biological target, so if they are conformationally mobile when free in solution, binding to the target causes a significant loss of rotational entropy of the compound, weakening the free energy of binding.

The frequent appearance of the number 5 in these empirical rules created the "Rule of 5" mnemonic for Lipinski's Rules.

Additional reading

Xue, L.; Bajorath, J. Molecular descriptors in chemoinformatics, computational combinatorial chemistry, and virtual screening. *Comb. Chem. High Throughput Screening* **2000**, *3*, 363-372.

Spellmeyer, D. C.; Grootenhuis, P. D. J. Recent developments in molecular diversity. Computational approaches to combinatorial chemistry. *Annu. Rep. Med. Chem.* **1999**, *34*, 287-296.

Agrafiotis, D. K.; Myslik, J. C.; Salemme, F. R. Advances in diversity profiling and combinatorial series design. *Mol. Diversity* **1999**, *4*, 1-22.

Warr, W. A. Combinatorial chemistry and molecular diversity. An overview. *J. Chem. Inf. Comput. Sci.* **1997**, *37*, 134-140.

Lipinski, C. A.; Lombardo, F.; Dominy, B. W.; Feeney, P. J. Experimental and computational approaches to estimate solubility and permeability in drug discovery and development settings. *Adv. Drug Deliv. Rev.* **2001**, *46*, 3-26.

Problems

1. Refer to the reaction in Problem 11.4. Use it in a library synthesis that begins with only commercially available compounds. Use SciFinder or ChemACX to determine the number of commercially available building blocks that can replace the acid chloride in this reaction sequence. How large a library could be created using just these acid chlorides? Who is the only supplier for 3-(2-thienyl)acryloyl chloride?

2. Which of the following compounds does not obey Lipinski's Rules?

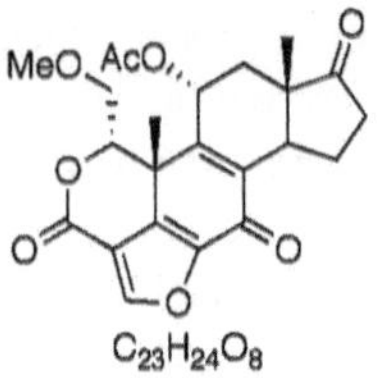

$C_{23}H_{24}O_8$ 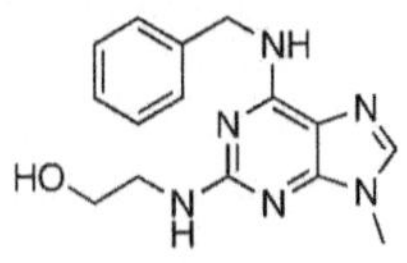 $C_{15}H_{18}N_6O$ 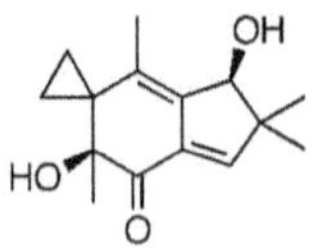 $C_{15}H_{20}O_3$ 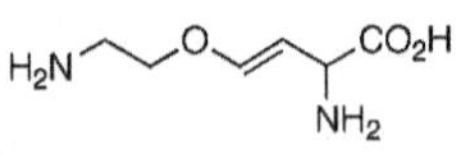 $C_6H_{12}N_2O_3$

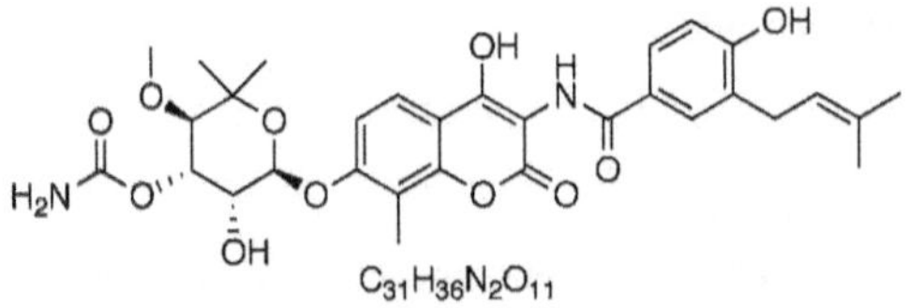

$C_{31}H_{36}N_2O_{11}$ 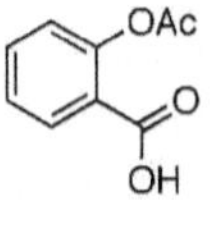 $C_9H_8O_4$

Nucleic Acid Microarrays

Miniaturization is one feature of combinatorial chemistry that enables large numbers of compounds to be prepared without making their handling unwieldy. Thus, solid-phase synthesis supports can carry single compounds on beads with dimensions of 10s-100s of microns (μm), a volume far smaller than any reaction vessel. Computer chips are a different field in which miniaturization has had a profound impact on the power of the technology. A method for the preparation of molecules using the same technologies utilized in semiconductor fabrication enables the generation of different molecules at different zones on a single surface. The resulting objects are now referred to as biochips or microarrays, and they have wide applications in biology. DNA chips, or nucleic acid microarrays, are the first commercial product to emerge directly from combinatorial chemistry.

The method of *light-directed synthesis* that was used to make the first peptide and DNA chips and is today used to make the most complex and powerful DNA chips was developed by the combination of two disparate techniques: photolithography and solid-phase synthesis. Photolithography is a method for forming patterns on a surface using a patterned mask (a "master," like a stencil) and a light-sensitive surface coating called a resist. In computer chip manufacturing, those patterns ultimately lead to the formation of semiconductors, conductors, and insulators to create transistors, memory elements, and microprocessors. Photolithography also enables many copies of a device to be made based on a single master. Solid-phase synthesis is well known to chemists working with peptides and DNA, and had even been applied in parallel fashion to monolithic substrates, such as in the peptides-on-pins method. However, the relationship of solid-phase synthesis to photolithography was not so obvious.

Light-directed synthesis was initially applied to the synthesis of arrays of over 1000 peptides and was demonstrated by mapping the epitope of the 3E7 anti-β-endorphin antibody. Each peptide sequence was held in a site 400 μm square. Binding of the antibody to the array was detected by fluorescence microscopy.

The extension of this initial peptide work to DNA was an important but also natural step. Solid-phase synthesis of DNA is a highly optimized technology that is considered the most effective and reliable method of chemical synthesis known. Another feature of DNA to be discussed later makes it the ideal molecule for use in light-directed synthesis.

As in any directed synthesis of an oligomeric molecule, during DNA synthesis one of the two hydroxyl groups of each bi-functional nucleoside monomer is protected with a protecting group (PG) so that uncontrolled polymerization does not occur. In the example compound on the following page, the 5'-hydroxyl group is protected, which is more common. Another protecting group (PG') is needed on the NH$_2$ groups of three of the heterocyclic bases (A, C, and G). The other (3') hydroxyl group of the nucleoside is activated for bond formation as a phosphoramidite group. The "amidite" also contains the phosphorous atom that will constitute the phosphodiester in the ultimate DNA molecule. Synthesis begins with removal of the protecting group on the solid phase. The resulting hydroxyl group then couples with the phosphoramidite under the mild acid catalysis of tetrazole. This creates a phosphite that is oxidized by iodine water to the phosphate, completing one round of synthesis. This process is repeated with different amidites to build up a desired sequence. "Gene machines" perform the synthesis by pumping reagents through a cartridge containing the support. After completion of the synthesis, the PG' group and β-cyanoethyl groups are removed from the DNA and it is detached from the surface under basic conditions. The result is a single-stranded DNA molecule.

Building blocks and chemistry of DNA synthesis

Two variations are necessary to adapt conventional DNA synthesis to light-directed synthesis. The first makes the attachment to the surface permanent. Light-directed synthesis goes to great trouble to synthesize molecules at specific locations on the surface. Information about the sequence would be lost if the DNA was detached from the surface. Further, one of the great virtues of DNA arrays is that many different sequences are attached to the *same* surface, permitting binding reactions of many molecules to be studied simultaneously in parallel.

The second variance from conventional DNA synthesis practice is in the protecting group on the 5'-hydroxyl of the monomer. As its name implies, light-directed synthesis requires chemistry that is promoted by light. More specifically, it requires that the sites on a surface at which chemistry occurs to be directed by light. Thus, one of the two main steps (deprotection or coupling) of the DNA synthesis cycle must be performed with light. The coupling of the support-bound hydroxyl group to the amidite is a very refined process, and it would be difficult to adapt this step to light control. The deprotection reaction, however, is a step that is readily adapted to light control through a large, well-known class of protecting groups that are photochemically removable. The photochemically cleavable linkers that have been discussed earlier in this book are examples of the effective use of light to promote chemical reactions in combinatorial chemistry.

The overall protocol for this example of light directed synthesis, then, is to spatially deprotect hydroxyl groups on the surface. The whole surface can be exposed to coupling reagents in the following step, but only sites that were addressed by light in the previous step will be coupled. This strategy is one of two that could be used to spatially direct the synthesis of an oligomeric molecule. The other strategy would be to deprotect the whole surface and direct the subsequent coupling only to specific sites. As exemplified on the facing page, either approach gives a similar microarray.

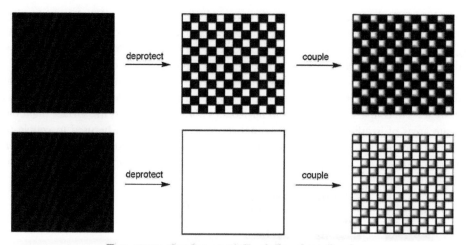

Two strategies for spatially-defined synthesis

Light-directed DNA synthesis begins with a glass surface coated with hydroxyl groups. These groups are protected with an X group that is photochemically removable. The surface is exposed to a light source through a mask M_1. The X groups are removed from the sites on the surface exposed to light. A coupling reagent X-A is washed across the whole surface, reacting only with sites that were addressed by light in the preceding step. The surface returns to a state

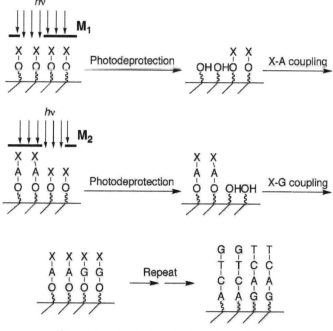

Steps in a light-directed DNA synthesis

in which each site bears a photochemically removable group because each building block also bears the photochemically removable X group. The surface is exposed to light through mask **M₂**, and washed with the coupling reagent X-G. Different sequences have thus been prepared in different sites. These processes are repeated to build up desired sequences.

The combinatorial element of light-directed synthesis is based on the ability to irradiate multiple sites on the surface simultaneously using the masks. Consider a set of stripe masks that are used to couple the four different nucleotides horizontally across the surface (light bars indicate areas of irradiation). An orthogonal set of stripe masks can be used to couple the nucleotides vertically across the surface. The result is an array of sixteen dinucleotides. The combinatorial nature of this array preparation is evident in the 16 sequences that were prepared in 8 chemical steps.

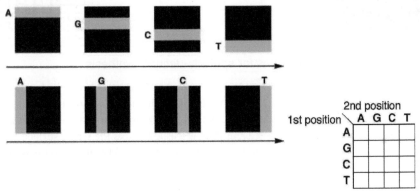

Dinucleotide array preparation

More generally, the number of sequences prepared in a light-directed synthesis is N^l, where N is the number of building blocks and l is the length of the oligomer synthesized. The number of steps in such a synthesis is $N \times l$. From calculations with these formulae emerge the particular advantages of light-directed synthesis for DNA array preparation, as compared to peptides.

$$\text{For DNA, } N = 4$$

$$\text{If } N \times l = 40, \text{ then } l = 10 \text{ and } N^l \approx 10^6$$

$$\text{For peptides, } N = 20$$

$$\text{If } N \times l = 40, \text{ then } l = 2 \text{ and } N^l = 400$$

The practical limitation on any synthesis is the number of steps, since they must be performed in sequence. Limiting each synthesis to 40 steps, peptide arrays of only two units can be prepared, giving a library of 400 sequences. With DNA, arrays of decanucleotides can be prepared, giving over a million sequences.

DNA arrays can be thought of as arising from combinatorial synthesis using spatial encoding: the identity of compounds is known based on their position on a surface. This format is also very apt for the use of the DNA once it has been synthesized. Traditional methods in molecular biology use oligonucleotides while they are connected to a solid phase, just as they are on a DNA array. Many conventional uses of surface-bound DNA involve hybridization, that is, formation of a double-stranded DNA from a single strand. Hybridization is driven by the energetic preference for double-stranded DNA, due both to hydrogen bonding between complementary bases across the double helix and stacking of aromatic rings along the double strand. The affinity of one strand for another is a function of both of these factors, but the specificity of double-stranded DNA formation is dependent on base pairing. Terms that are used to describe a hybridization event in the array format are *probes*, the synthetic DNA sequences on the array, and *target*, the DNA(s) in solution that will be hybridizing to one or more of the probes.

Consider a site on a surface bearing the sequence CTGACTGA. It may hybridize to its perfect complement with the formation of eight Watson-Crick base pairs, or it may hybridize to a sequence with a different base at one position. Because only seven base pairs are formed, the binding of this latter sequence will not be as strong and the equilibrium constant for hybrid formation will be less favorable. In many cases it is possible to arrange conditions in a hybridization experiment (called stringency) such that only the binding of perfect complements to surface sites is favored, whereas binding of DNAs that contain mismatches at even one base are disfavored.

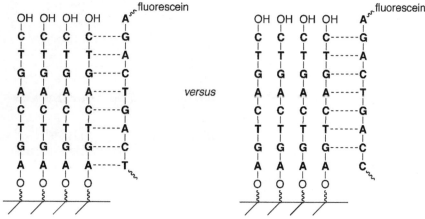

Hybridization of surface-attached DNA.

Microarrays are very powerful tools when used in hybridizations because the identity of each synthesized sequence is known based on its position, so the identity of sequences that will bind at a particular position can be determined on the basis of the base-pairing rules. Most often, the DNA that is bound to the array carries some sort of optical label so that it can be visualized by microscopy or a related imaging technique. The structure of an unknown DNA molecule is determined by a chain of information that leads all the way back to the sequence of chemical reactions that were performed at each position on the array. Microarrays are even more powerful because of the large number of *different* sequences that can be prepared on an array, due to the combinatorial nature of the synthesis.

A second example shows a primitive DNA microarray that was actually prepared. The sites on the surface that initiate synthesis are protected with photochemically removable groups. The specific group used in commercial DNA microarray production is a nitrobenzyl group, the MeNPoc group, attached to the surface by a linker molecule

Linker-MeNPoc derivatized glass surface

The microarray was prepared with 4^4 probes (256) of the sequence CG⊗⊗⊗⊗CG in 16 chemical steps. The physical layout of the array depicted below shows that masks are related recursively. This microarray was hybridized with one oligonucleotide target of sequence GCGGCGGC that also incorporated a fluorescent group. The array was then imaged by fluorescence microscopy. A large fluorescence signal at the position of the complementary sequence is seen in the micrograph.

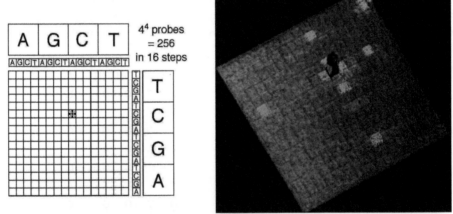

Probe layout for an 8-mer array and its hybridization to a fluorescent DNA target
Reprinted with permission from *Proc. Natl Acad. Sci. USA* **1994**, *91*, 5022-5026.
Copyright 1994 National Academy of Sciences, U.S.A.

The power of current commercial DNA arrays made by Affymetrix goes far beyond this early example. The sizes of the probe sites is 25 μm square for routine production. The length of the probes is 25 nucleotides. The substrate on which the probes are held is a 1.4 cm × 1.4 cm piece of fused quartz that is diced from a larger substrate on which many copies of an array are synthesized. Each array is mounted in a plastic cartridge that permits the addition of a solution of the DNA target and subsequent washing solutions through two ports. The array is visualized through a glass window.

Arrays are designed with DNA sequences chosen for many purposes. Arrays often address sequences within a specific organism (e.g., yeast, *E. coli*, mouse). Generally, multiple probes are included for each sequence to be certain its presence is reliably identified. In new developments, such as the HG-U133 array that represents a large fraction of the human genome, the probe site size has been reduced to 18 μm square, making accessible over 600,000 probes on each array. These probes address ~33,000 known human genes and ~39,000 cDNA transcripts using >45,000 sets of probes (multiple probes are used for each target).

Applications of DNA arrays made by combinatorial synthesis include: DNA sequencing, mutation detection/disease diagnostics, research applications like physical mapping of genes, and molecular computation. One of the most popular uses for DNA arrays is in gene expression analysis. Its purpose is determination of all of the genes that are being expressed in a cell (and at what level) at a particular time, and particularly in comparison between cells that may have different histories. While gene expression ultimately leads to the production of proteins, which are the actors in the cell, gene expression analysis takes a snapshot of the genes as their mRNA as they make their way from DNA to protein. The RNA is converted to cDNA for the hybridization step. The huge numbers of probes available on DNA arrays makes possible the assessment of the expression level of essentially every gene in a cell. Such a global gene expression study is an experiment that could not have even been conceived before the development of DNA arrays.

With these DNA microarrays established, other workers have sought to make microarrays using different methods. One simple method to make them is to physically deliver DNA to small spots and arrange its attachment to the surface. The methods for physical delivery primarily involve a liquid jet, such as is used in printers, or a stylus. These methods can make spots of 100 μm or perhaps a little less. However, they require pre-existing DNA, and would not be considered a combinatorial synthesis. Many biology laboratories do have huge collections of DNA sequences relevant to their particular research interests, so these spotting methods offer them very practical access to microarrays.

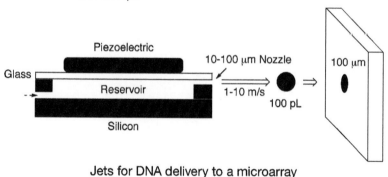

Jets for DNA delivery to a microarray

Concepts related to light-directed synthesis can be used in combinatorial DNA microarray fabrication. As was earlier discussed, either the deprotection step or the coupling step in each DNA synthesis cycle can be spatially directed to combinatorially prepare an array. By deprotecting hydroxyl groups across the whole surface of the array but delivering only one amidite to each site, different sequences are made at different sites. This method relies on a barrier of a fluoropolymer between each of the synthesis sites to prevent amidite solutions from mixing. It also relies on physical methods of reagent delivery, such as jets, and thus can prepare probe sites of about 100 μm. This method is currently applied commercially by Agilent.

Additional reading

Pirrung, M. C. How to make a DNA chip, *Angew. Chem. Int. Ed.* **2002**, *41*, 1276.

Problems

1. How many DNA sequences are prepared in how many steps in a light-directed microarray synthesis of 12-mers?

2. A novel building block set of nucleic acid-like monomers is shown below. These compounds are used to assemble 10-mers that will be used to bind to DNA. How many distinct DNA sequences could be targeted by this type of compound with a perfect sequence match? (Hint: be sure to write the full structure of at least a couple of repeating units and compare it to DNA structure).

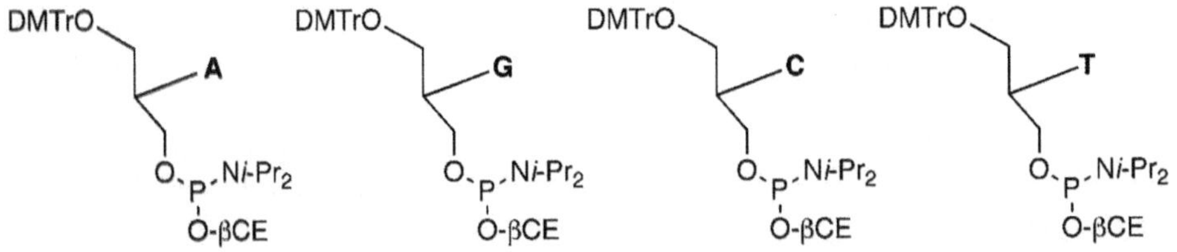

DMTrO   A
O P N*i*-Pr₂
O-βCE

DMTrO   G
O P N*i*-Pr₂
O-βCE

DMTrO   C
O P N*i*-Pr₂
O-βCE

DMTrO   T
O P N*i*-Pr₂
O-βCE

Combinatorial Materials Chemistry

Miniaturization and parallel processing can have as large an impact on the development of novel materials as it has had on drug development. The techniques will necessarily be different, as solid-phase synthesis on polymer beads is not relevant to materials. The standard format that has evolved for developing new materials is a two-dimensional array where the composition of the material is a function of its position. These libraries look much like DNA arrays.

Materials chemistry also tends to assemble elements rather than molecules. Elements can be deposited by "printing" methods, vapor deposition or dipping into a reagent solution. Vapor deposition techniques include electron-beam and thermal evaporation, sputtering, and pulsed laser ablation. Solutions of elements in their ionic forms can be used in the dipping methods, but during a curing phase following the synthesis, elements, oxides, or alloys may be formed.

A shadow masking method for deposition of metal oxide thin films is very similar to classic stencils. Overlaid on a single crystal $LaAlO_3$ substrate is a primary mask that forms 128 1 mm × 2 mm synthesis regions in a 16 × 8 format.

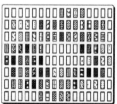

Array of magnetic materials

An ingenious method is used to create combinations of different elements in every region. The binary masking approach was borrowed from the initial studies on peptide arrays. An example of a binary masking sequence is shown on the following page. Binary masks are defined as masks that expose half of the substrate in each step of the synthesis. Each mask overlaps by half with the mask from the preceding step. In this example, four building blocks A, B, C, and D will be deposited in one corner of a 4 × 4 array, and all possible deletions from the building block set ABCD will be placed in the other 15 zones.

A 16-member binary library of superconductors was formed from Bi, Pb, Ca, and Sr precursors, in 1:1 stoichiometry, with Cu added to all locations. Secondary masks that fit the binary synthesis protocol were overlaid on the primary mask. Precursors were sputtered onto the surface through the secondary masks in a stepwise fashion, thereby generating an array containing all 2^n combinations that can be formed by deletions from the series of n masking-deposition steps. The stoichiometry of the phases can be modified by varying the sputtering time to vary the film thickness of each deposition. After coating, the substrate is subjected to thermal and

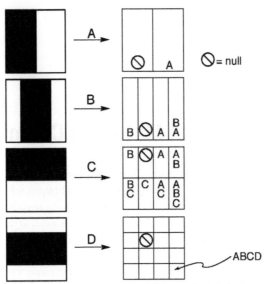

Binary masking makes ABCD and all deletions

oxidative treatment to form the final array of phases. The resistance of each site was measured by a scanning SQUID magnetometer as a function of temperature, and superconductivity (80–90 K T_c) was found in two sites: known high-temperature superconductors $BiCuCaSrO_x$, and $BiPbCuCaSrO_x$. The dependence of the properties of the $BiCuCaSrO_x$ film on stoichiometry and deposition order was investigated in a 128-member library, and a $Bi_2Sr_2Ca_2Cu_3O_{10}$ phase with even higher T_c (110 K) was found.

    Another masking strategy, called quaternary, uses each mask in the deposition of up to four different precursors, with a 90° rotation between depositions. Quaternary masking leads to much wider variations in composition than binary "deletion" libraries. Like combinatorial DNA synthesis, quaternary masking can generate 4^n compositions in $4n$ steps.

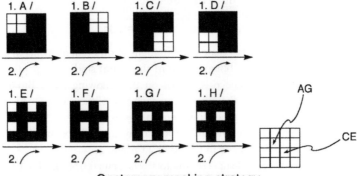

Quaternary masking strategy

    These shadow masking techniques were applied to the creation of libraries of some 25,000 compositions of V, Ba, $BaTiO_3$, Sr, Sr_2CO_3, TiO_3, Al_2O_3, MgO, SnO_2, Y_2O_3, Eu_2O_3, La_2O_3,

Tb_4O_7, Tm_2O_3, and CeO_2 that were tested as phosphors. The image below shows a red phosphor, $Y_nAl_nLa_nV_nO_n$ doped with Eu, discovered from this effort.

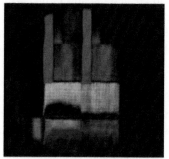

Phosphor library

Reprinted with permission from *Angew. Chem. Int. Ed.* **1999**, *38*, 2494 - 2532
Copyright 1999 WILEY-VCH.

Another class of materials that were discovered using such methods are x-ray phosphors.

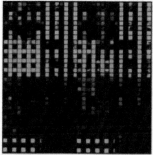

A library of 1024 phosphors (GdGaO on SiO_2)

Reprinted with permission from *Science* **1998**, *279*, 1712-1714
Copyright 1998 AAAS.

As in all combinatorial methods, the identification of a hit in an initial screen of materials must be followed up. Hits can be verified by re-synthesis and testing, and the composition and/or structure must be determined. Unlike the situation with DNA arrays or molecular encoding, the composition at a particular position in a materials array may not be known, though the conditions used to produce the substance at that position are known. A larger sample of an interesting thin film could be readily prepared by conventional methods (i.e., not involving spatial segregation). The characterization of thin films can still be challenging, and ideally would not require re-synthesis. X-ray methods are broadly useful for determining the structures of materials, and have been adapted to interrogate locations as small as 2 μm in thin films. X-ray microprobe methods include X-ray fluorescence, X-ray diffraction, and near-edge X-ray absorption fine structure spectroscopy (EXAFS). These methods do require a synchrotron beam-line, however.

Another variant of the mask-based approach to creating different compositions across a surface involves the formation of layers of varying thickness by moving the mask during deposition. Some zones are exposed to greater amounts of precursor, some to less. This technique was applied to optimize an organic light-emitting diode (LED). These devices have an emitter layer of a fluorescent metal hydroxyquinolinate, a hole transport layer (HTL) of triphenyldiamines, and may have a hole-blocking electron transport layer (HBETL) of spiroquinoxaline. Surprisingly, these compounds are sufficiently volatile to be evaporated onto an indium tin oxide (ITO) electrode substrate as a mask is moved across it. The goal of this work was discovery of the optimum thickness of each layer. Deposition could be controlled from 50-150 nm.

The ink-jet approach to delivering reagents to particular points on a surface that was used in DNA chip fabrication can also be used to deliver solutions of material precursors if they are not sufficiently volatile.

Solution-based strategies can also permit the generation of gradients that lead to different compositions at different points on a surface. The compositional spread technique deposits three precursors at the corners of a triangle and generates a physical embodiment of the ternary

Step 1: evaporation of HTL with constant thickness (40 nm) on ITO-substrate

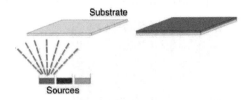

Step 2: preparation of Alq$_3$-gradient (0 - 130 nm) by vapor deposition and simultaneous shutter movement

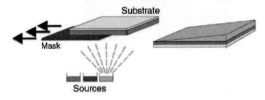

Step 3: rotation of substrate by 90° and evaporation of HBETL (0 - 30 nm) by simultaneous shutter movement

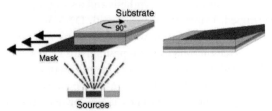

Gradient deposition method for organic LEDs

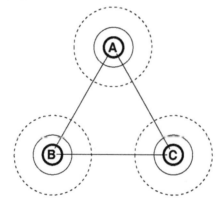

aluminum hydroxyquinolinate
(Alq₃)

spiroquinoxaline

Emitter molecule and hole-blocking electron transport molecule for organic LEDs

phase diagram by diffusion. Scanning for a desired property generates the final dimension of the phase diagram. Through the preparation of about 30 libraries of dielectric materials using this method, a novel material $Zr_{0.15}Sn_{0.3}Ti_{0.55}O_2$ was identified that is superior to amorphous silicon dioxide as an insulator.

Ternary compositions generated by diffusion

If the kinetics of the deposition of a metal species from solution are relatively slow, linear and step gradients of a substance can be formed by dipping a substrate into a solution of its precursor at a controlled rate. Colloidal gold fits this requirement, enabling silver-clad gold particles of varying compositions to be formed on thiolated glass. By turning the substrate at right angles and repeating the process with a different metal, a range of compositions in two dimensions can be generated. Similar to the mask-based strategies, the composition at each site may not be immediately known, but the conditions required to generate it are, and the phase can be reproduced. Using a stepper motor with 3 μm resolution, it is possible to create by dipping >10^8 different compositions a 3 cm × 3 cm substrate.

This gradient deposition method was applied to discover an optimum silver-clad gold surface for surface-enhanced Raman scattering (SERS). On a 2 cm × 2 cm substrate, the gold was deposited over 6 h, while the silver ion was deposited over 22 min. A SERS instrument with

2 mm spatial resolution was used to scan the surface to locate the most active region within a dynamic range of 1000, which was further probed to locate a region over 100-fold more effective than background.

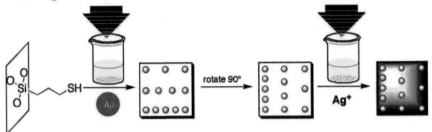

Dipping method for nanoparticle formation

Additional reading

Jandeleit, B.; Schaefer, D. J.; Powers, T. S.; Turner, H. W.; Weinberg, W. H. Combinatorial materials science and catalysis. *Angew. Chem. Int. Ed.* **1999**, *38*, 2494-2532.

Problems

1. A quaternary masking strategy is used to make binary mixtures of indium, tin, silicon, and germanium chlorides. After high temperature annealing and oxidation, how many different combinations of semiconductor electrodes will be generated?

Combinatorial Catalyst Discovery

The application of a technique as powerful as combinatorial chemistry to a field of chemistry as important as catalysis is natural. Catalysis is a broad area, encompassing both heterogeneous and homogeneous catalysts. Some combinatorial catalysis methods resemble the methods used for small molecule drug discovery, mostly in homogeneous catalysis, while others look more like methods in combinatorial materials chemistry, mostly in heterogeneous catalysis.

A significant challenge in this field is detecting the library members that are good catalysts. Unlike many of the other properties that have been sought in libraries, the ability to catalyze reactions is not a property for which general assays exist. It is straightforward to develop an assay for catalysis of one specific reaction, but to perform such an assay on beads or in a high-throughput manner can be difficult, or at least expensive and slow. Even more challenging is detecting catalysts that prefer one enantiomer over the other, the essence of asymmetric catalysis.

There is one general property of catalyzed reactions - they are exothermic. The catalyst only affects the rate of the reaction, so the heat released per unit time is a reflection of the catalyst efficiency. It might seem challenging to detect the heat evolved from a reaction, but in fact that is exactly what a calorimeter does (though it does so in a thermally isolated environment). The preferred small scale of combinatorial chemistry challenges the sensitivity of calorimeters, but modern microcalorimeters are capable of detecting even weak molecular complex formation and would be capable of observing catalysis on a small scale. However, high-throughput calorimetry is probably a concept that will (justifiably) never be tested. A method for *in situ* (on arrays or beads) detection of the heat evolved in a reaction is needed, but the need to perform such experiments without significant thermal shielding makes stringent demands on current technologies.

A solution to this problem is infrared (IR) thermography, a technology better known as "night vision". Very sensitive IR cameras detect the heat evolved from catalysts promoting reactions at point sources, and can resolve temperature differences as small as 0.02 °C. Those catalysts can be of known composition based on their location, or of a composition determinable by methods earlier covered.

A simple example shows the application of this method to catalyst discovery. It is well known that 4-dialkylaminopyridines are nucleophilic catalysts for acylation reactions. They act via formation of the N-acylpyridinium ion that is a more reactive acylating agent than the activated acid from which it is derived.

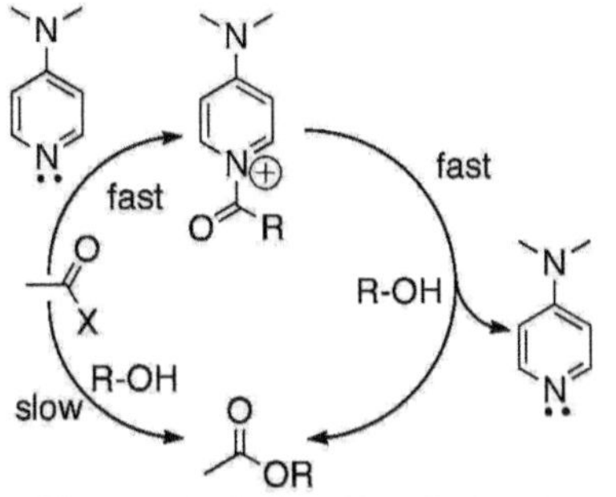

Nucleophilic catalysis by dimethylaminopyridine

The dialkylaminopyridine structure was included in a peptide/peptoid library of 3150 compounds made by split/couple/mix synthesis that was encoded using the Still method. The beads were placed in an isopycnic solution containing acetic anhydride and ethanol. Beads that are active catalysts warm from the heat of their reaction and can be detected by the IR camera. An example of such an experiment is shown on the facing page. "Hot" beads are selected and decoded to identify the active catalyst structure. Resynthesis of these compounds in quantity and testing on solid phase enabled their catalytic activity to be verified, with the quinuclidine below exhibiting the strongest catalysis. The frequency with which a particular coded compound was found in the library correlated with its catalytic activity when it was prepared as a single pure compound.

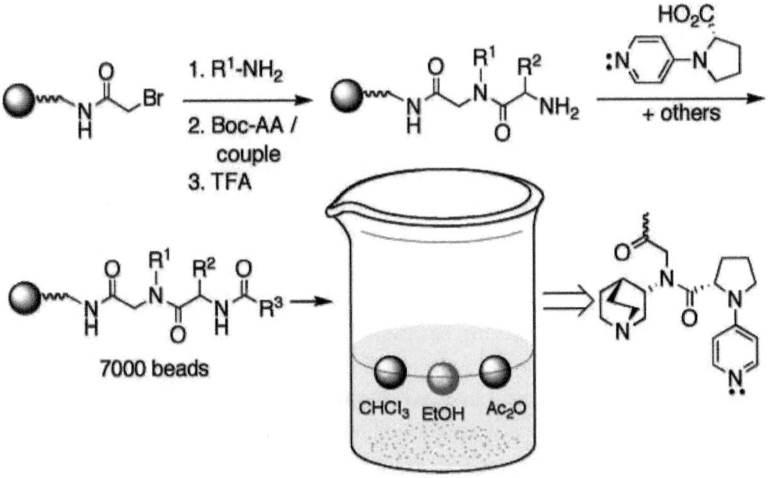

Library of acylation catalysts

IR thermographic techniques can also be used in the evaluation of heterogeneous catalysts. Transition metal-on-metal oxide catalysts were generated by deposition of precursors in ~200 µg amounts. An array of 37 such catalysts was imaged in the presence of 3-hexyne and hydrogen using IR thermography. Active catalysts for alkyne hydrogenation were 1-5 mol % Pd on SiO_2.

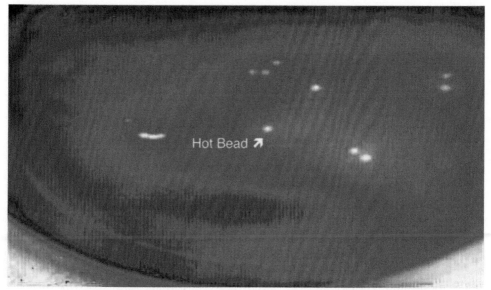

Thermographic imaging of the catalyst library

The deposition from solution of materials precursors delivered by jets was discussed in the preceding section. Likewise, catalyst precursors can be spatially deposited by jet in many combinations for optimization of catalyst properties. For example, catalysts for the direct conversion of methanol in a fuel cell (anode electrocatalysts) were discovered. Transition metal chloride salt inks of Pt, Os, Rh, Ir, and Ru were printed on carbon paper and reduced with borohydride to metalize them. The carbon paper support also served as an electrode. Imaging of the catalytic power of the compositions was enabled by the recognition that the oxidative half-cell reaction produces acid, so a pH-sensitive fluorescent dye can signal regions catalyzing the desired reaction. In this case, quinine was used as the indicator. In the presence of quinine/methanol solutions, those compositions converting methanol to CO_2 will also produce acid and thereby promote local fluorescence. Images are shown on the following page.

$$CH_3OH + H_2O \rightarrow CO_2 + 6\ H^+ + 6\ e^-$$

$$O_2 + 4\ e^- + 4\ H^+ \rightarrow 2\ H_2O$$

Over 600 different compositions were examined in this library. An unusual composition (44% Pt / 41% Ru / 10% Os / 5% Ir) was discovered that has a far higher current density than previously known direct methanol electrocatalysts.

Ternary phases of methanol electrocatalysts

Catalysts for asymmetric reactions can also be studied using combinatorial methods. Many of these examples entail a metal-mediated desymmetrization process in which ligand-accelerated catalysis is observed. The powerful principle of ligand-accelerated catalysis, which is exhibited by many highly selective asymmetric processes, assures that reaction rates of the un-ligated, achiral catalysts (that would not be enantioselective) are slower than that of the asymmetric pathway. The measure of the effectiveness of an asymmetric catalyst is the enantiomeric excess of the product. In the reaction shown below, the enantiomeric excess (*ee*) is simply the percent of the major enantiomer less the percent of the minor enantiomer. If a reaction is selective in producing only one enantiomer, the *ee* is 100%.

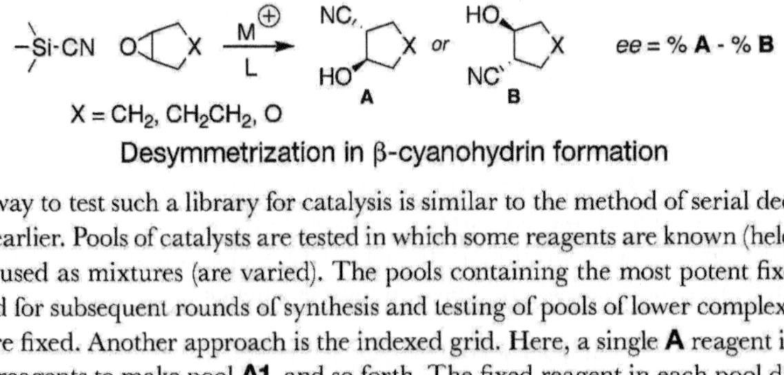

$$ee = \% \, \mathbf{A} - \% \, \mathbf{B}$$

$X = CH_2, CH_2CH_2, O$

Desymmetrization in β-cyanohydrin formation

One way to test such a library for catalysis is similar to the method of serial deconvolution described earlier. Pools of catalysts are tested in which some reagents are known (held fixed) and others are used as mixtures (are varied). The pools containing the most potent fixed reagents are selected for subsequent rounds of synthesis and testing of pools of lower complexity as other reagents are fixed. Another approach is the indexed grid. Here, a single **A** reagent is combined with all **B** reagents to make pool **A1**, and so forth. The fixed reagent in each pool distinguishes

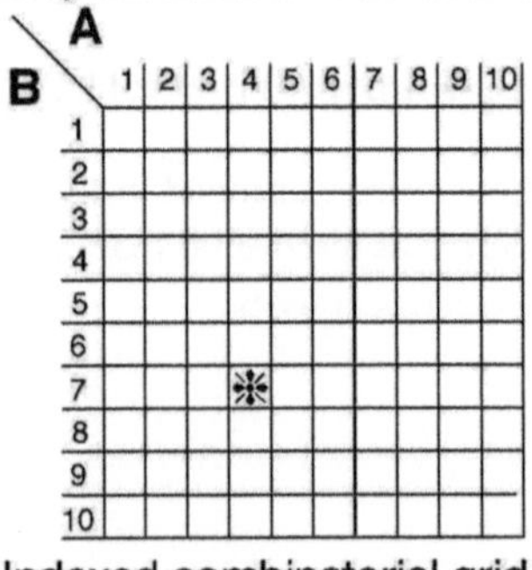

Indexed combinatorial grid

its activity from other pools, so identifying the most potent pools in each dimension should identify the best combination at their intersection.

An example of the use of combinatorial principles in asymmetric synthesis is the addition of Me_3SiCN to cyclohexene oxide, for which catalysis was studied with a family of peptide ligands bound to Lewis acids. The salicaldehyde imine (salen) functionality was included because titanium-salen complexes were known to catalyze achiral versions of the epoxide opening reaction. The basic structure of the library is a dipeptide condensed at its amino terminus with an o-hydroxybenzaldehyde, creating a chiral metal binding pocket. In the presence of a metal salt, a chiral complex should form on the solid phase that could catalyze a reaction in solution.

Salen-peptide-Lewis acid catalyst library

The ultimate purpose of this library was to discover an asymmetric catalyst. However, because screening for enantioselectivity often requires tedious chromatographic analyses, a step-wise approach was taken to select catalysts that would be later subjected to a more complete analysis of catalysis.

The first step was discovering ligand and metal combinations that exhibit ligand-accelerated catalysis. This was accomplished by preparing a mixture of all ligands and combining them with five individual metal isopropoxides. The titanium salt exhibited by far the greatest rate acceleration. Examining a mixture of all five metals with ten individual ligands identified two (derived from salicaldehyde) that were eventually shown by a serial procedure to be the best catalysts. This was not a major discovery since salens were already known to catalyze this reaction and they were the only two salens present in the library.

Ligand library

More interesting was the optimization of asymmetric catalysis, not just ligand-accelerated catalysis. The procedure used here has similarities to serial deconvolution. However, it does not involve mixed positions in early rounds of optimization, so it potentially could be vulnerable to the problem that the "best" molecule identified is dependent on the starting point. Schematically, as shown on the following page, the process involves moving sequentially and at right angles through a multi-dimensional property space of catalysts (here, three dimensions). At each step, the best catalyst is chosen as the starting point for the next generation.

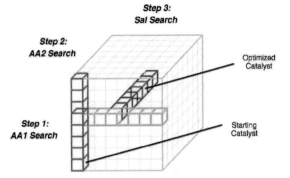

Search of a three-dimensional space for asymmetric catalysts using a serial method

The catalyst ultimately identified (~90% *ee*) was far superior to the first generation catalysts (20-55% *ee*, average 36% *ee*).

generation 1 (10 Rs)
R = *tert*-Bu, ~55% *ee*

generation 2 (16 Rs)
R = CH(CH₃)O-*tert*-Bu, ~65% *ee*

generation 3 (13 Rs)
R = *o*-F, ~90% *ee*

diversity = 10 × 16 × 13 = 2080

Serial determination of an effective catalyst

Another asymmetric reaction that was examined using similar parallel synthesis methods is the addition of cyanide to an imine to produce an α-aminonitrile, the classical Strecker reaction from amino acid synthesis. The design of the ligand library included an amino acid (AA) linked to a diamine (DA) linked to a salicaldehyde imine (Sal), somewhat like the metal-binding ligands discussed above. However, it was demonstrated in earlier libraries that the addition of metal salts *reduced* the enantiomeric excess in this addition reaction. Therefore, this asymmetric reaction is *not* metal-catalyzed. The capability of peptide derivatives to promote this reaction with enantioselectivity must be based on direct hydrogen bonding and other interactions with the reactants. Further investigations omitted metals.

AA—DA—Sal
11   3   4

1. catalyst /
NC-Si
2.
F₃C   O   CF₃

A library of catalysts for the Strecker reaction

The library compounds were prepared by solid-phase synthesis on polystyrene and examined as catalysts while attached to the support. The enantiomers of the trifluoroacetyl α-aminonitrile products are separable by gas chromatography (GC) on a chiral column, enabling the enantioselectivity of the reaction to be directly determined.

The design of this 132 compound library used hydrophobic amino acids, a thiourea linker to an (R,R)-diamine, and o-tert-butyl-salicaldehydes. The best catalyst in this library was synthesized in soluble form and examined in Strecker reactions of aromatic and bulky alkyl aldehydes. The products were obtained in quite good yield and often acceptable enantiomeric excess.

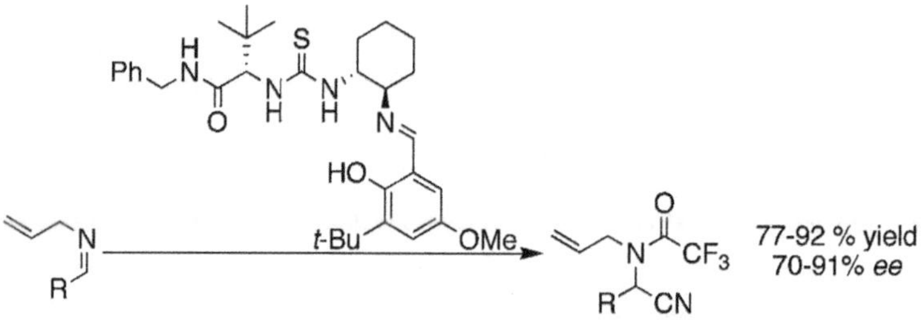

Catalytic activity of a library-derived compound

Note that in these two examples, the catalysts were examined while on solid phase or in solution, but they were necessarily examined one at a time. As was seen with the acylation catalysts described earlier, a method to examine catalysts in parallel while on the solid phase (where they were prepared) is a more powerful technology.

Such an improvement has been made by the incorporation of an indicator of catalytic activity into the solid-phase synthesis or assay. It is based on fluorescent compounds whose fluorescence is dependent on the presence of protons. In anthracene-amine compounds, the fluorescence intrinsic to the anthracene itself is quenched by electron transfer from the amine. However, if the amine is protonated, making its unshared electrons unavailable for quenching, fluorescence returns. One can see how this process could be readily engineered into an indictor of any reaction that produces acid. This detection scheme is not as general as the thermographic imaging of the heat of a reaction, but is still fairly broad. In the example below, an acylation reaction produces acetic acid, which activates the fluorescence of the anthracene.

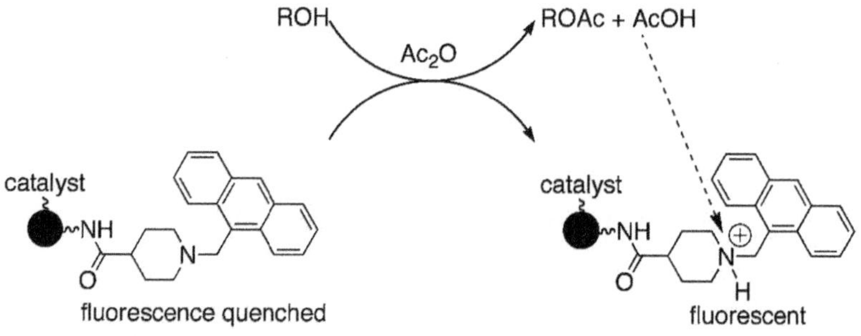

Catalytic activity detected by proton-enhanced fluorescence

This method was applied to the discovery of an enantioselective acylation catalyst. In this reaction, a racemic alcohol is (ideally) converted into an ester derived from one enantiomer of the alcohol, and the other enantiomer of the alcohol remains unreacted. Many catalysts for such reactions are known, almost all of them being enzymes called lipases. Each of these enzymes has its own intrinsic substrate preferences, efficiency, and selectivity of acylation. There may be many reactants for which no good enzymatic catalyst exists. A general approach to discovery of such catalysts for specific reactants would therefore be very valuable.

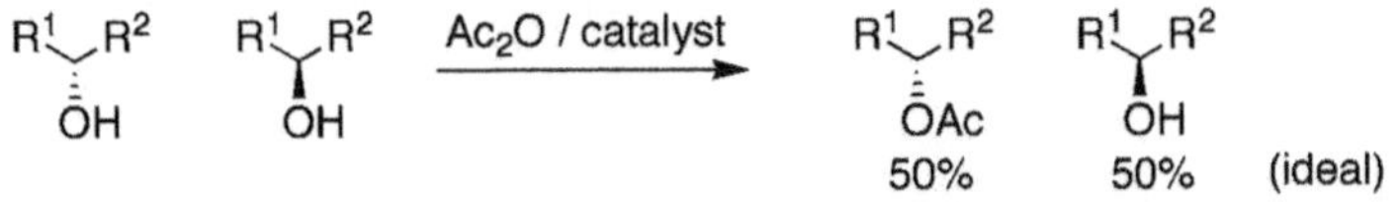

Enantioselective acylation (kinetic resolution)

Like the dimethylaminopyridine-catalyzed reaction discussed earlier, this acylation reaction is subject to nucleophilic catalysis. Another nucleophilic catalyst is N-methylimidazole. A chiral building block for peptide synthesis that is an analog of N-methylimidazole can be readily prepared from the amino acid histidine. This building block can then be incorporated into a peptide library that can be examined for the ability to catalyze a desired acylation reaction. Since the peptide is a chiral molecule, it will have a different rate of reaction with each of the two enantiomers of a racemic alcohol and may thereby exhibit enantioselectivity.

The design of the peptide library placed N-methylhistidine at the N-terminus and alanine at the C-terminus, with one of fourteen genetically coded amino acids at six intervening positions. The calculated diversity of this library is 14^6, or 7.5 million, but the actual number of beads used in the synthesis was only 100,000, so every possible peptide catalyst was not represented. These peptides were synthesized on 500 μm support beads bearing the anthracene-amine to indicate catalysis.

The screening of this library was performed by dissolving an alcohol of interest in toluene and adding acetic anhydride and ~1000 beads of the library. After one hour, the brightest fluorescent beads were selected and reagents were washed from them. They were individually submitted to the same reaction conditions and the rate of fluorescence development was observed. The fastest catalytic beads were subjected to Edman sequence analysis. One peptide that catalyzes the kinetic resolution of 1-phenylethanol was identified in this way. The peptide was then resynthesized as its methyl ester shown below.

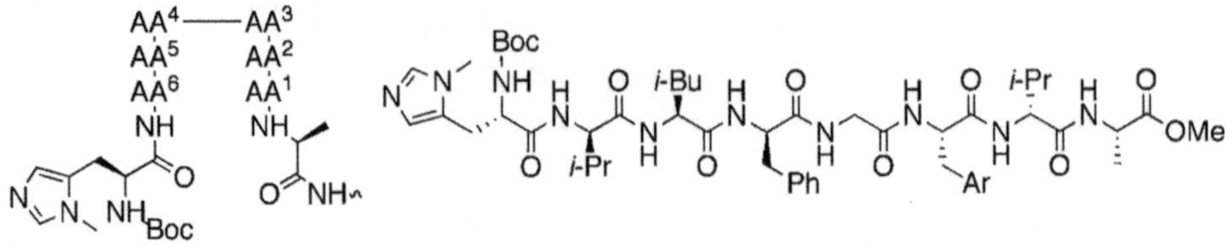

Selected peptide kinetic resolution catalysts

Several other peptides discovered in this library were examined for their ability to promote the enantioselective acylation of 1-phenylethanol. Interestingly, some selectively favor the acylation of one enantiomer of the alcohol, while others favor acylation of the other enantiomer.

Clearly, this trait could not have been designed in a group of catalysts and could only have been discovered by a diversity-based approach.

A second generation, 6,000 peptide library was generated based on modifications of the best of these sequences. Testing of this library was focused on direct determination of the enantioselectivity. This could only be accomplished with single beads, and was performed with one bead in each well of a 96-well plate. An optimum peptide catalyst was discovered and its structure was determined as before. As the methyl ester, it acylates 1-phenylethanol with 20:1 selectivity. Related molecules such as 1-naphthylethanol are acylated by this catalyst with selectivities as high as 50:1.

Enantioselective acylation of 1-phenylethanol

The necessity of incorporating the proton-enhanced fluorophore into the solid-phase synthesis makes this method less general than it might be. An alternative was developed by incorporation of the proton-enhanced fluorophore into a polymer gel in which the screening would be conducted. The gel serves to limit the diffusion of acid from the vicinity of active catalyst beads. They can be identified from the fluorescence surrounding them and manually selected.

Preparation of a proton-sensitive fluorescent gel

The gel was based on a known polymer called PEGA. The anthracene-amine was added into its synthesis to incorporate it into the polymer gel. When acidified, this gel becomes fluorescent, and it can be cast around the beads of a catalyst library. The acid generated during the acylation reaction can promote the fluorescence of the nearby gel. Evidence of catalysis is seen in a halo of fluorescence around the bead.

As can be appreciated from the foregoing, the determination of the enantiomeric excess of the product of an asymmetric reaction can become problematic when it must be performed either in high-throughput form for catalysts prepared in parallel synthesis, or particularly when the catalysts are immobilized on solid-phase synthesis beads. One ingenious approach to this problem has been reported. Mass spectrometry of course cannot distinguish between enantiomers, as they have the same mass, but it is possible to label one enantiomer of a pair with a heavy isotope to form a mixture called quasi-enantiomers. If the reaction progress is followed by MS, the rate of disappearance of the two isotope peaks will be different if the catalyst exhibits enantioselectivity.

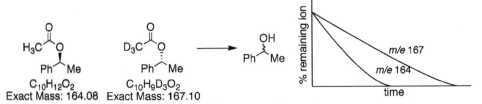

Quasi-enantiomers for MS-based determination of enantioselectivity

Additional reading

Jandeleit, B.; Schaefer, D. J.; Powers, T. S.; Turner, H. W.; Weinberg, W. H. Combinatorial materials science and catalysis. *Angew. Chem. Int. Ed.* **1999**, *38*, 2494-2532.

Senkan, S. Combinatorial heterogeneous catalysis-a new path in an old field. *Angew. Chem. Int. Ed.* **2001**, *40*, 312-329.

Reetz, M. T. Combinatorial and evolution-based methods in the creation of enantioselective catalysts. *Angew. Chem. Int. Ed.* **2001**, *40*, 284-310.

Taylor, S. J.; Morken, J. P. Thermographic selection and effective catalysts from an encoded polymer-bound library. *Science* **1998**, *280*, 267.

Reddington, E.; Sapienza, A.; Gurau, B.; Viswanathan, R.; Sarangapani, S.; Smotkin, E. S.; Mallouk, T. E. Combinatorial electrochemistry: a highly parallel, optical screening method for discovery of better electrocatalysts. *Science* **1998**, *280*, 1735.

Copeland, G. T.; Miller, S. J. Selection of enantioselective acyl transfer catalysts from a pooled peptide library through a fluorescence-based activity assay: an approach to kinetic resolution of secondary alcohols of broad structural scope. *J. Am. Chem. Soc.* **2001**, *123*, 6496.

Problems

1. Why is ligand-accelerated catalysis important in reactions creating optically active compounds?

2. What is the most general property of catalyzed reactions? This property suggests what method to detect catalysis?

3. Describe how a ligand-accelerated catalyst could be discovered for the following reaction using the indexed grid method .

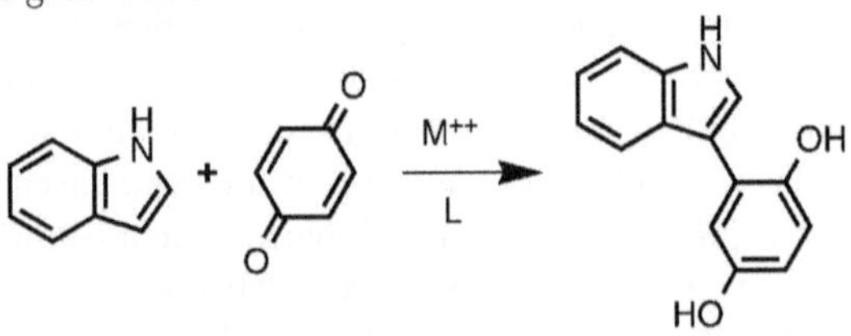

Peptides on Phage

The concept of phage display arose first as a result of the same challenge that prompted Geysen to develop his peptides-on-pins method: determining the sequence of a protein that is recognized by an antibody. Smith therefore called his biologically generated peptides an "epitope library." The key to this method is the central dogma of molecular biology, which allows a synthetic DNA sequence to be translated into a peptide sequence on the surface of a bacteriophage. It also represented an important conceptual leap because in phage display, the molecule that has binding activity can be directly related (because it is physically attached) to a molecule that encodes its structure. Encoding techniques have become widely used and very important in combinatorial chemistry. In order to understand how this method works, it is necessary first to understand the structure and life cycle of a bacteriophage.

Bacteriophages are viruses that infect bacteria. While there are many different classes of phage, one in particular has been widely used in molecular biology. The filamentous class is exemplified by M13, which is often used in DNA sequencing and site-directed mutagenesis. M13 has the interesting feature that its genome is only a single strand of DNA, instead of a double strand. In order for an *E. coli* cell to be infected by M13, it must bear an extra-chromosomal genetic element (like a plasmid) called the F-factor. This causes the production of a structure on the outside of the cell called the F-pilus. The F-pilus provides a route of infection by M13 and other filamentous phage.

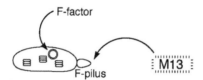

Infection of *E. coli* by M13

Once inside the cell, the single-stranded phage DNA is converted to a double-stranded form by the host cell DNA polymerase. This double-stranded replicative form (RF) is then replicated to the level of about 100 copies per cell. Transcription and translation from this double-stranded plasmid produce the proteins that will coat the phage particle. This is a particularly important event in phage peptide display, as it is the stage at which the library members are created. A special replication process produces copies of the single-stranded phage genome, which is packaged in the coat proteins and extruded through the cell membrane of the bacterium. When these bacteria are grown on agar plates, the phage-infected bacteria grow more slowly, creating a plaque.

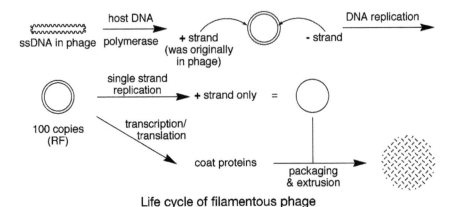

Life cycle of filamentous phage

The DNA in the RF of these phage encodes ten proteins, five of which envelop the single-stranded genome of the mature phage particles. The proteins are given Roman numerals. The most common is pVIII, which stacks up along the length of the phage. As many copies of pVIII as are needed to cover the DNA are incorporated during the assembly of the viral particle, enabling relatively large additions of DNA to the phage genome. At the phage tip are five copies of the pIII protein, which is involved in recognition of the F-pilus and phage infection. However, it happens that changes can be made in the sequence of pIII without affecting the ability of the phage to infect *E. coli* too severely. Genetic engineering of pIII has been used in most phage display of short peptide sequences. It may be desirable to have only a few copies of the peptide library members on each phage so that the binding of the phage to the target is not too tight.

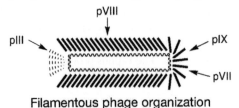

Filamentous phage organization

The strategy to create a phage peptide library is to make a DNA sequence that encodes random peptides and insert it into the phage DNA so that random peptide sequences are present on the surface of the phage. If a specific sequence on the surface is recognized by the target, the phage bearing it can be separated from those that are not, and then its DNA can be sequenced to find out the sequence of the peptide that has the binding activity.

The first step in creating a phage display peptide library is the creation of DNA that encodes the peptide. This library-encoding oligonucleotide should have at each base triplet a mixture of the codons for each of the amino acids that are to be included in the library. Recall that the genetic code is degenerate, with some amino acids being coded by multiple base triplets. Furthermore, among the choices for those that are degenerate, the specific codon(s) that is/are used for a particular amino acid varies between different organisms. This is a function not just of the DNA sequence, of course, but also of the sequences in the anti-codon loops of the transfer RNAs. Codon usage must be correct to efficiently create a genetically engineered protein.

An ideal method for creating oligonucleotides encoding a peptide library would use building blocks for DNA synthesis that are the codons used in *E. coli*. However, it is not very practical to synthesize and use these 20 compounds, so the alternative of designing the DNA based on the degeneracy of the genetic code has been taken. A mixture of all four bases is used at the first two positions of each triplet. The third position has a mixture of dG and either dC or dT. These mixtures are designated NNK and NNM. They encode each of the 20 amino acids but only one stop codon (you can verify this for yourself by comparing this design to the genetic code given in the first section). These DNA sequences can be prepared by automated oligonucleotide synthesis. They are also synthesized with flanking regions that enable them to be inserted into a specific restriction enzyme cleavage site within the RF of the phage. A popular site for insertion is in the pIII protein, whose *N*-terminal end bears a signal peptide that is cleaved as the phage is packaged and extruded from the cell. When the random DNA is inserted into the phage DNA between the sequence encoding the signal peptide and the gene 3 sequence, the random peptides are expressed on the *N*-terminus of the pIII protein. All of the pIII proteins on an individual phage will bear the *same* peptide insertion at the *N*-terminus, but different phage in the library will have different inserted peptide sequences. Peptides on phage can therefore be considered one phage/one compound libraries.

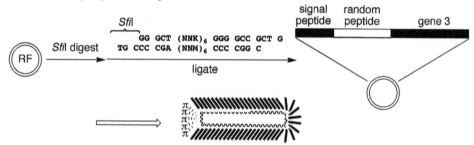

Creation of phage expressing random peptides (shown as π)

The creation of the actual DNA construct that is inserted into the phage DNA is not simple. It must be double-stranded, but synthetic DNA molecules are obtained as single strands. A known sequence is therefore added at the 3'-end that binds a primer for DNA polymerase. This enzyme extends the primer, adding the complements to whatever bases are found at the randomized positions in each template.

```
NNN NNN NNN NGG GCT (NNK)₆ GGG GCC GCT GXX XXX XXX XXX    DNA polymerase
                                         CYY YYY YYY YYY   ──────────────►
                                                              dNTPs

  Sfil digest          GG GCT (NNK)₆ GGG GCC GCT G
──────────────►        TG CCC CGA (NNM)₆ CCC CGG C
```

Generation of dsDNA insert

The M13 RF library created by the ligation must be transfected into *E. coli* to enable phage replication. It is important that transfection is efficient so that a large number of independent recombinant phages are created. If the random DNA insert encodes hexapeptides, the number of possible peptide sequences will be 20^6, or 64 million. If it encodes decapeptides, $\sim 10^{13}$ peptides are possible. Somewhat similar to the situation regarding the number of beads and the calculated diversity in a split/couple/mix synthesis, the number of independent recombinants in the phage library must far exceed the calculated diversity of the combinatorial library if all of the combinations are statistically likely to be present. For example, with the 1.3×10^9 heptapeptides, 90% of all sequences will be present when there are 7.9×10^{10} transformants. If the computed library size is less than the number of transformants, some sequences will be missing. Techniques for efficient transformation include electroporation, which can produce libraries of $\sim 10^9$ transformants/μg of vector DNA.

A typical library contains 6-15 randomized residues. Libraries of 7-mer and 12-mer random peptides inserted into pIII are commercially available. Other features that might be incorporated into the design of the peptide library are conformational constraints. Beginning and ending the sequence with cysteine should create a peptide family of cyclic molecules linked by a disulfide.

Linear and cyclic octapeptide library designs

The screening of a phage peptide library, sometimes called "panning," requires that the molecule for which a peptide ligand is sought be immobilized on a solid phase. For large molecules like proteins, this might be as simple as non-specific adsorption to a plastic or glass petri dish. For some small molecules, a biotin molecule might be added so that the ensemble can be bound to a protein such as avidin that does adsorb to surfaces. Direct covalent attachment can also be used.

The process of enrichment of the phage library for those with a desired binding activity is somewhat like panning for gold. Weakly bound phage (the equivalent of sand) are sloshed out of a tray, while those that are tightly bound (the equivalent of gold nuggets) are retained.

In order to eliminate phage that merely like to bind to the petri dish, or those that bind to any immobilized protein by a non-specific interaction, an initial depletion process can be performed. In the example on the facing page, bovine serum albumin (BSA) is used to provide non-specific binding sites. Those phage that do not bind non-specifically are used in a second round of panning for binding to the specific target. Those phage that did not bind in the BSA experiment but also do not bind specifically to the target are washed out in this step. By these two rounds of depletion from the library, those phage that do not have a specific binding interaction with the target are eliminated, and those that do have a specific interaction are immobilized. By a change of conditions, such as pH or salt concentration, the tightly bound phage can be eluted into solution. This family of phage has presumably been enriched in those that express peptides that bind to the immobilized target and must have a significantly enhanced affinity for the target as compared to the initial, diverse peptide library. This phage population may represent only a small fraction of the number of phage with which panning was begun - indeed, that is the purpose of the process. This enriched phage fraction is used to infect *E. coli* and a new generation of phage particles is produced to provide sufficient phage to perform another cycle of selection. This step is called amplification. Further rounds of enrichment should lead to a phage population that is dominated by those displaying sequences with the strongest binding to the target.

Panning can also be made more selective by including a competitor for the specific binding site on the target during the panning or elution process. This can help insure that the selected phage not only bind to the target, they bind in competition with other molecules that bind to a specific site on the target.

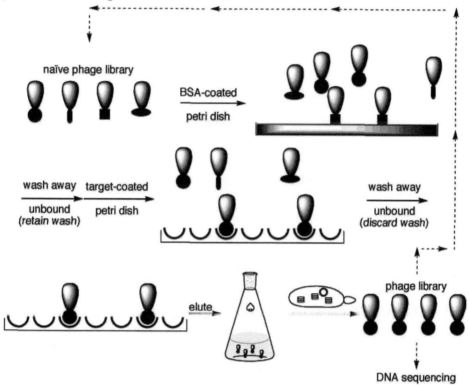

Panning cycles enrich a population of phage for those displaying peptides with specific binding activity toward an immobilized target

An alternative to the panning procedure is affinity column chromatography, which is shown on the following page. The target is attached to a chromatographic support and eluted with a phage library solution. Those phage with a weak affinity for the support will have high mobility, will elute quickly, and will be discarded. Those with a high affinity for the support will have low mobility and will be retained. It might even be necessary to include denaturants in the eluting solution to get the most tightly bound phage particles to elute. This situation would be ideal. After phage enriched in binding abilities are selected, the phage are used to infect *E. coli*, DNA from individual clones is sequenced, and the sequence of the binding peptides can be inferred. This is the general concept of phage display, but specific aspects of the method are best learned through examples.

Affinity chromatography-enriched phage library

A wide range of peptide ligands for protein targets have been discovered through phage display, some of which are given in the Table 4.

Table 4. Proteins for which peptide ligands were discovered by phage display

Target	Discovered sequence	K_D (μM)
Streptavidin	CHPQFC	0.27
Concanavalin A	DVFYPYPYASGS	46
GPIIb/IIIa	Ac-CRGDMFGC-NH$_2$	3.5
SH3 domain (Src)	KGGGAAPPLPPRNRPRL	0.24
SH3 domain (Yes)	VSLARRPLPPLP	0.67
IL-1 receptor	FEWTPGYWQPYALPL	0.0019
Urokinase receptor	AEPMPHSLNFSQYLWYT	0.01

Applications of phage display

One of the best known examples of the success of phage display is the discovery of a peptide that mimics the actions of the cytokine erythropoietin (EPO), a therapeutic for anemia. EPO induces the proliferation of bone marrow via a trans-membrane protein receptor that is a symmetric dimer. As a protein, EPO must be prepared by recombinant DNA techniques and administered by injection. A small molecule that accomplishes what EPO does would constitute a very important advance.

A cyclic disulfide library was created in phage with 12 positions of diversity and was panned against an immobilized version of the extra-cellular binding domain of the EPO receptor. After several rounds of panning and amplification, families of peptide sequences were identified. Their consensus is YXCXXGPXTWXCXP, where X represents positions that can be occupied by several different amino acids. None of these sequences is represented in the primary structure of EPO, so the peptide that was discovered did not merely identify the binding domain of the protein, but rather identified a molecule that truly mimics the protein. Surprisingly, not only does this peptide bind to the receptor, it activates it, and it exhibits many of the biological effects of EPO in cell-based assays and in mice.

Chemical synthesis of one of these peptide sequences EMP-1 (EPO-mimetic peptide 1, GGTY̱SCHFGPLTW̱VCKPQGG) and examination of a co-crystal structure with the binding domain of the EPO receptor showed that the peptide forms a non-covalent dimeric complex and bridges two receptor proteins. Important in this interaction are the Y and W underlined in the sequence above.

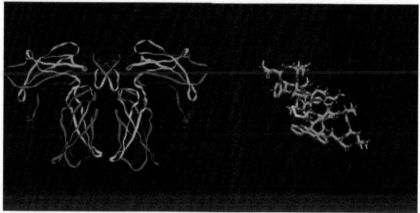

L: Quaternary complex of EPO receptor with EMP-1; R: Bound peptide conformation

Synthesis of covalently linked, dimeric peptides led to ligands that are over 100× as potent. The basis of this effect is as follows. The binding of each of two different peptide molecules to the receptor must be enthalpically favorable, but is entropically unfavorable because of the loss of the translational entropy of each peptide. Linking them *before* the binding event means that the enthalpic contribution from binding of the dimeric peptide should be about twice that of each individual peptide, but only a single entropy penalty must be paid upon binding.

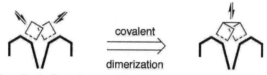

Covalent dimeric peptides have increased potency

One of the features of phage display that is potentially problematic is that even if peptides with ideal binding properties are discovered, they are necessarily formed from (S)-amino acids. They are therefore subject to degradation *in vivo* by proteases and would likely not make very good therapeutics. Ideally, one would like to discover a peptide composed of all (R)-amino acids because they are resistant to proteases, but there are no enantiomeric phage from which to select tight binding peptides. However, an alternative is available based on screening a phage library against a target that is enantiomeric to the actual *in vivo* target. A discovered sequence of all (*S*)-peptides that binds to an all (*R*)-receptor must be identical to the sequence of all (*R*)-peptides that binds to an all (*S*)-receptor.

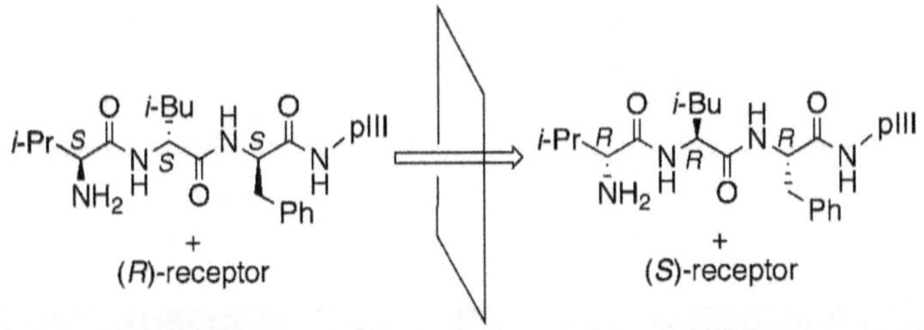

Peptides discovered by phage display against an enantiomeric receptor identify (*R*)-peptides that will bind to the natural receptor

The challenge of this method is access to the enantiomeric receptor. It is not available from Nature, of course, but in many cases is available by chemical synthesis. Major advances have been made in the total chemical synthesis of proteins, and polypeptides of hundreds of amino acids can be prepared by expert laboratories.

This method was applied to the discovery of ligands for Src homology 3 (SH3) domains. This is a 50-residue domain present in many cellular proteins that mediate intracellular signal transduction. They are important for cell growth, migration, differentiation, and responses to the external environment. The enantiomeric SH3 domain was prepared by chemical synthesis and used to pan a 14-residue cyclic disulfide peptide library. The sequence identified was RCLSGLRLGLVPCA. This sequence was then synthesized using all (*R*)-amino acids and examined for its interaction with the native SH3 domain. It exhibited a 63 µM dissociation constant. Compared to the 0.24 µM dissociation constant measured for the all (*S*)-peptide listed in Table 4, perhaps this experiment was not an unqualified success, but it did establish an important principle.

Additional reading

Wrighton, N. C.; Farrell, F. X.; Chang, R.; Kashyap, A. K.; Barbone, F. P.; Mulcahy, L. S.; Johnson, D. L.; Barrett, R. W.; Jolliffe, L. K, Dower, W. J. Small peptides as potent mimetics of the protein hormone erythropoietin. *Science* **1996**, *273*, 458-64.

Schumacher, T. N.; Mayr, L. M.; Minor, D. L., Jr.; Milhollen, M. A.; Burgess, M. W.; Kim, P. S. Identification of D-peptide ligands through mirror-image phage display. *Science* **1996**, *271*, 1854.

Problems

1. A phage display peptide library experiment is performed to find a sequence that binds specifically to maltose. A (poly)maltose adsorbent column is prepared. A filamentous phage library is prepared with an insertion of 10 random codons into the gene for the pIII protein. After transformation of *E. coli* with the engineered replicative form, 10^9 independent recombinant phage particles (transformants) are produced. This library is loaded onto the (poly)maltose column and eluted, with the amount of phage eluting observed by ultraviolet absorbance. The last-eluting 0.1% of the original library is isolated and used to transform *E. coli*, but no plaques are obtained. What do you think is wrong in this experiment, and how could it be rescued?

2. A random phage peptide library was panned against the (100) face of gallium arsenide (GaAs). The selected phage were used to infect *E. coli* and amplified by 10^6. This process was repeated 5 times. Eluted phage were used to infect *E. coli*, clones were isolated, and DNA of each was sequenced – the sequence of one such clone is GUC ACU AGU CCU GAU AGC ACC ACC GGA GCU AUG GCU. What is the sequence of the peptide that binds to GaAs(100)? Suggest a mode of binding.

3. Why is mixture synthesis rather than split/couple/mix synthesis used to create the random DNA sequences for phage peptide libraries?

4. If a library of peptides on phage were eluted through an ATP affinity column, which part of the elution profile shown should be collected to obtain the sequences that bind to ATP?

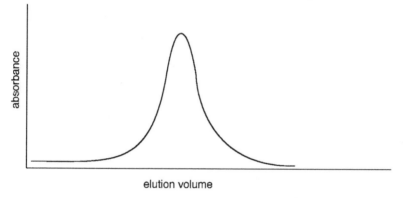

elution volume

Nucleic Acid Selection

The process of nucleic acid selection has almost no scientific predecessors. Developed around 1990, it exploits the well-known capability of nucleic acids for self-replication, but adds a previously unappreciated property: specific binding. Like phage display, it moves through a number of generations of selection, with amplification following each generation. One term applied to the process is the acronym SELEX, standing for systematic evolution of ligands by exponential amplification. Given the term "amplification," a method of molecular self-replication is inherent to the process, and generally involves polymerase chain reaction (PCR). The goal of a SELEX experiment is to discover a nucleic acid sequence from among a large library of sequences that has specific binding activity to a particular target. Such sequences are called *aptamers*. More sophisticated goals, such as catalysis of specific biochemical reactions, can also be sought in more complex experiments. The method will be illustrated with the discovery of binding molecules that are RNA. RNA has a much more diverse secondary structure than DNA, particularly in the single-stranded form.

A library of DNA sequences serves as an archive of diversity that is used to initiate SELEX experiments. The creation of the library is performed by chemical synthesis using the isokinetic mixture method, as was done in the creation of the coding sequences for phage display of peptides. This synthetic library can be copied for use in individual binding experiments. To enable production of an RNA copy of that DNA library, a specific sequence must also be present that permits the binding/recognition by an RNA polymerase. These enzymes require such *promoter* regions. RNA polymerase performs *transcription* of DNA to RNA beginning at the promoter. This simply means a processive synthesis from the 5'-end toward the 3'-end. A means for amplification of those RNA molecules that are selected is also needed. Known, constant sequences are added flanking the synthetic library region to serve as binding sequences for primers in a PCR amplification.

The structure of a nucleic acid pool created for use in SELEX is given below. Beginning at the 5'-end of this single-stranded molecule, a promoter sequence for a specific RNA polymerase that is commonly used to transcribe RNA, T7 RNA polymerase, is added. This region can also be the binding site for a PCR primer in later steps. The random sequence region might be of any length, but typically is less than 100 nucleotides. This number still creates tremendous

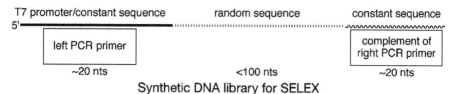

Synthetic DNA library for SELEX

sequence diversity in the library, up to 4^{100}. Because 4^{100} is $\sim 10^{60}$ and the number of molecules that can actually be prepared in such an experiment is far less (recall that Avogadro's Number is $\sim 10^{24}$), most of the molecules that can be imputed from the diversity in a nucleic acid library are not in fact present.

The execution of a SELEX experiment with such a library as the starting point is performed as follows. The synthetic DNA library is amplified by a few cycles of PCR to make a double-stranded DNA pool. In the presence of ribose nucleotide triphosphates (NTPs), the T7 RNA polymerase enzyme makes an RNA copy of one of these two DNA strands, including the random bases. The specific binding sequence to be discovered is within these bases. It is practical at this stage to have created $\sim 10^{15}$ different RNA molecules. The actual sampling of the imputed diversity of the library would be complete only if the length of the random sequence region were < 25, because 4^{25} is $\sim 10^{15}$.

This single-stranded RNA pool is then subjected to binding experiments that separate those molecules that bind specifically to the target from those that do not. This concept is familiar from phage display, and examples will be given later. The RNA that results from this initial fractionation is by design present in very small amounts, since only a few of the sequences are expected to bind to the target, and only the most potent binders should be taken into the next generation. With few desired sequences being present at low levels, larger quantities of this nucleic acid are needed to permit the execution of another round of selection without concern for mechanical and other losses. RNA is not itself self-replicating, but it can be converted to a complementary copy called cDNA by the enzyme reverse transcriptase (RT). RT makes a single-stranded DNA copy of an RNA template using deoxyribose nucleotide triphosphates (dNTPs). The ssDNA that results from this process is very similar to the synthetic DNA pool that was used to begin SELEX, with the exception that it has been enriched in sequences in the random region that code RNAs that can specifically bind to the target. The flanking sequences that enable PCR remain intact, so this cDNA can be used to create many copies of the binding sequences in a new double-stranded DNA pool. This completes one cycle of selection and

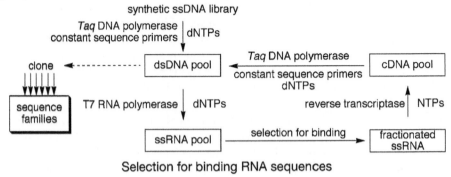

Selection for binding RNA sequences

After several of these cycles have been executed, the sequence of the DNA in this pool is determined, which corresponds to the sequence of the RNA that binds to the target. This dsDNA pool cannot be directly sequenced, since it is still likely to be a mixture at many of the positions in the random region, and only the average sequences of the binding molecules could be obtained by sequencing the conglomerate. To identify representative RNA sequences, it is necessary to obtain sequence information on a random selection of the *individual* DNA molecules in this pool. This can only be accomplished by cloning them into bacterial vectors for amplification and isolating DNA from single bacterial colonies for sequencing.

A specific example gives form to this general scheme. An ATP-agarose is available for purification of ATP-binding proteins, but can also be used in a chromatographic method for isolation of ATP-binding RNA molecules. A pool of ~10^{14} different RNA molecules was prepared with 72 random nucleotides flanked by the defined sequences. A large quantity of RNA (10 copies of each sequence) was used in the initial selection from the naïve library to ensure that as many as possible of the imputed sequences are present and to maximize the opportunity to discover a sequence with the desired binding activity. The nucleic acid was denatured before loading onto the column, and was allowed to equilibrate before elution. The non-specifically bound sequences were eluted with column buffer and the specifically bound sequences were eluted using a high concentration of ATP. The nucleic acid eluting in this second fraction was amplified by RT-PCR and a new RNA population was generated by transcription. After six cycles of selection and amplification, ATP-binding sequences came to dominate the pool. The DNA was cloned and the sequences were determined.

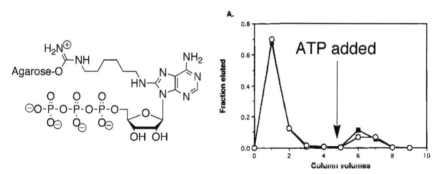

ATP matrix used in RNA selection by affinity chromatography

Reprinted with permission from *Biochemistry* **1995**, *34*,
656-665. Copyright 1995 American Chemical Society.

A 40-mer sequence discovered from this experiment is given below. Using bioinformatics tools, it can be predicted to adopt the secondary structure shown.

GGG UUG GGA AGA AAC UGU GGC ACU UCG GUG CCA GCA ACC C

```
              G A A
            A       A
          A           C
          G           U
            G   G     U
      GGGUUG   UGGCAC   U
      ||||||   ||||||
      CCCAAC   ACCGUG   C
            G        G
```

The structure of this RNA was determined by NMR methods. The ATP is bound to the RNA aptamer through a G•A mismatch. Its purine ring is intercalated between the purine rings of A10 and G11 in the loop. Its K_D for ATP was measured to be ~1 μM.

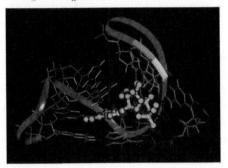

Structure of the aptamer-ATP complex

Printed with permission, D. Burke, Indiana University.

Interestingly, the *same* aptamer sequence is found by selections with immobilized NAD$^+$ and S-adenosyl methionine. It is clear that the key feature of these molecules that is being recognized by RNA libraries is the adenosine.

NAD$^+$ and S-adenosyl methionine

In another example, a selective binder to riboflavin was selected. A non-derivatized pre-column was used to select out those molecules that bind to the support matrix or to the linker used to attach the specific ligand to the support. The initial RNA had a complexity of 5×10^{14} unique sequences. After five rounds of enrichment, 10% of the RNA bound to the matrix and was eluted with riboflavin. This enriched pool was still complex (10^6 molecules), so further selections against non-specific binders were used. Sequencing of clones following round nine gave G-rich repeats. The binding of this sequence to riboflavin was dependent on potassium ion, suggesting that the riboflavin binding motif has a G-quartet structure, which is favored by potassium ion. This aptamer is shown on the facing page.

A number of chromatographic supports bearing dyes are available for protein purification. One such support was used to discover aptamers that specifically bind malachite green and other related dyes using SELEX. The 3-dimensional structure of the complex of the aptamer with malachite green was determined. This structure is presented in both framework and space-filling forms on the facing page. The RNA has a significant superstructure that is involved not in direct interactions with the dye, but in establishing the overall binding motif. Just a few atoms (in red) of the malachite green are visible in the center of the space-filling model.

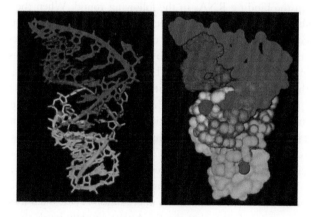

Sequences of riboflavin aptamers, a riboflavin affinity matrix, the proposed aptamer dual G-quartet structure, and G-quartet stabilization by potassium ion

Framework and space-filling models of an aptamer-malachite green complex

The classes of molecules that can be recognized by RNAs derived from SELEX experiments include nucleotides, nucleosides, amino acids, enzymatic cofactors, and basic antibiotics such as aminoglycosides. It has even been possible to find an RNA that could distinguish theophylline and caffeine, which differ by only a methyl group, with selectivity $> 10^4$.

Malachite green, theophylline and caffeine

Table 5. Aptamers to Small Molecules

Molecule	K_D (μM)
Adenosine	1
Theophylline	0.11
Arginine	0.33
Citrulline	62
Tryptophan	18
Vitamin B12	0.09
Flavin	0.5
Tobramycin	0.0008
NAD	2.5

SELEX is an example of directed evolution toward a goal. Biological evolution includes several characteristics such as natural genetic diversity, genetic recombination, biological selection, and mutation. SELEX as already described incorporates diversity and selection, and can incorporate *mutation*. The PCR reaction that is used to amplify RNA following the selection step can be performed so that base changes are introduced. The resulting dsDNA pool causes sequence changes in the binding RNAs in the following round. When the number of actual molecules in the library used in the initial selection is smaller than its imputed diversity, this strategy permits access to more of the potential diversity in binding sequences.

Aptamers have also been selected to bind to protein targets. In some ways, they are equivalent to antibodies in terms of providing specific binding reagents for proteins, and can be much easier to generate than antibodies. Their affinities have been comparable, 50 nM to 50 pM. For example, a 27 nt aptamer has been identified that binds to vascular endothelial growth factor (VEGF). It constitutes an interesting therapeutic called Macugen™ that is in clinical development for treatment of age-related macular degeneration.

A potential problem in using aptamers *in vivo*, either in cells or as therapeutics, is their sensitivity to nucleic acid degrading enzymes. However, modifications of nucleic acids that increase stability to such enzymes are known. Including modified bases in the RNA derived from a SELEX experiment should make it resistant to degradation *in vivo*. Such modifications can be exploited in one of two ways. Given that the RNA that actually binds the target molecule is formed exclusively from NTPs in the RNA polymerase-catalyzed reaction, the native version of a base can be replaced with a modified base. However, the RNA polymerase must accept the non-native NTP and the reverse transcriptase must accept the corresponding non-native nucleotide in the template strand. Examples of modified NTPs include the 2'-fluoronucleotides given on the facing page. As these novel bases have unique functional groups compared to the native bases, they are incorporated in place of native bases from the first round of selection when possible. This is called front-loaded SELEX. Nuclease-resistant bases that are *not* accepted by the nucleic acid polymerases used in SELEX can only be incorporated once strongly binding aptamers have been identified. For example, 2'-O-Me purines increase nuclease resistance and may be substituted for the purines in an aptamer. Aptamers that include 2'-O-Me purines must necessarily be prepared by chemical synthesis. The effect of the modification on binding must then be examined.

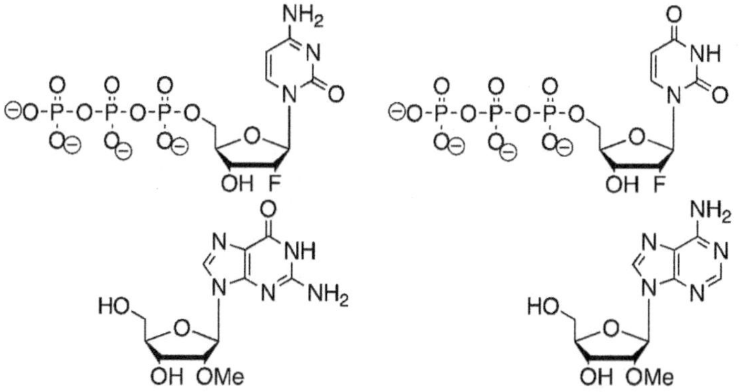

NTPs and nucleosides that stabilize aptamers

Structural investigation of aptamers has been very fruitful. They demonstrate a wide variety of energetically accessible secondary structures, including loops and hairpins.

In select cases, it has been possible to create biased systems in which RNA molecules that perform certain reactions can be selected from a pool. However, these have primarily been specialized reactions; novel, desirable, and *general* catalytic activities have not yet been discovered from nucleic acid libraries.

Using methods related to those discussed above, it is also possible to create libraries of DNA molecules and select them for binding and catalytic activities. Interestingly, when DNA and RNA are selected for the same binding partner, there is generally no sequence relatedness between the binders from the two different polymers.

Additional reading

Wilson, D. S.; Szostak, J. W. In vitro selection of functional nucleic acids. *Annu. Rev. Biochem.* **1999**, *68*, 611.

Gold, L. Globular oligonucleotide screening via the SELEX process: Aptamers as high-affinity, high specificity compounds for drug development and proteomic diagnostics, in "Combinatorial Chemistry And Technology: Principles, Methods, And Applications," S. Miertus, G. Fassina (Eds.), Marcel Dekker (1999), p. 389.

Problems

1. To create an RNA library bearing only the nuclease-resistant 2'-fluoropyrimidines, a random region of DNA of length 40 was synthesized using only pyrimidines. How many possible binding sequences will be present in this library? If the SELEX process is begun with 10^{15} sequences, how long could the random region be while still ensuring full coverage of the computed diversity?

2. *L*-Cocaine, the enantiomer of the naturally occurring drug of abuse, was prepared by total synthesis. It was immobilized onto a dextrose column through its nitrogen atom. A DNA aptamer library with 30 randomized positions was created by PCR, enzymatically converted

to its single-stranded form, and passed down the column. The last eluting 0.001% of the library was isolated. This DNA sample was again subjected to PCR, conversion to single-strand, and affinity chromatography. After several repetitions of this process, the DNA was put into a bacterial cloning vector for sequencing. The consensus sequence emerging from this experiment is shown. Note that it does not reflect all the possible positions of diversity in the library. Describe how you would prepare a compound based on this discovered sequence if you wished to use it to treat cocaine addiction by binding to the drug and preventing its actions.

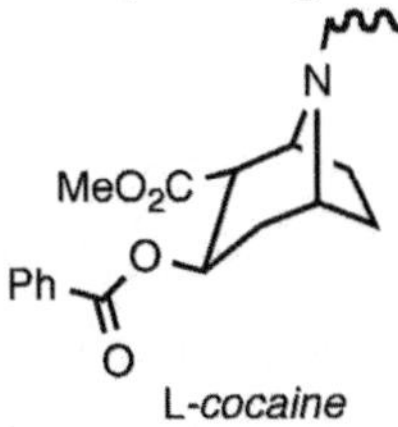

MeO$_2$C

Ph

5'-AGTCTTAGTCATCTAGCTA-3'

L-*cocaine*

3. A SELEX experiment is performed using 4-thioUTP in place of UTP in creating the RNA. Your challenge is to find a sequence that will bind to Hg^{++} but not Ag$^+$. Describe a process by which you would enrich the library in molecules with this property.

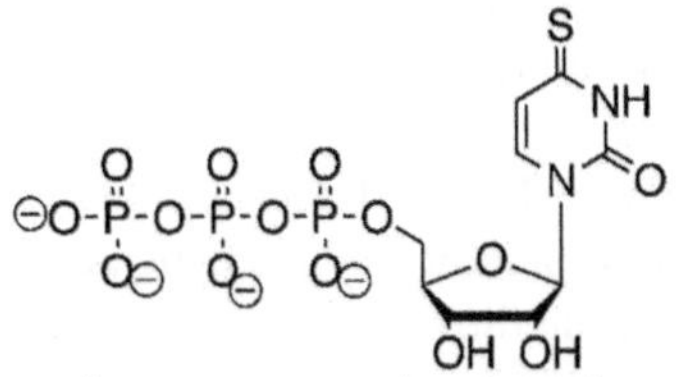

$\ominus$O-P-O-P-O-P-O

OH OH

4. The following affinity matrix was prepared and used to adsorb a random RNA library. After washing, what would be a good molecule to specifically elute the tight-binding RNA molecules?

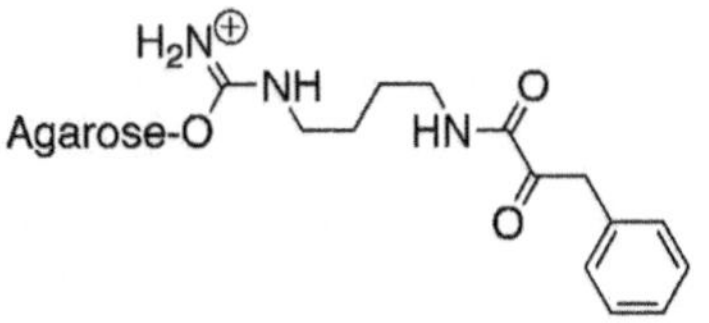

H$_2$N$^\oplus$

-NH

Agarose-O   HN

Complex Combinatorial and Solid-phase Synthesis

While many of the advances in combinatorial chemistry principles were made with highly reliable chemistries, such as peptides, they do not offer true chemical diversity, as we have earlier discussed. However, many chemists have focused on improving this situation, and it could be said today that most reactions in organic chemistry can be conducted in combinatorial form, either in solution or on solid phase. There are many ways to approach the synthesis of sophisticated chemical structures in library format.

Epibatidine

We will begin this section with solution-phase syntheses. Ley's concept of a synthesis in which all reagents are in solid-phase form was applied to the amphibian natural product epibatidine, which has analgesic activity via a novel target, the nicotinic acetylcholine receptor. The bridged structure of epibatidine was proposed to derive from an internal substitution reaction of a cyclohexylamine. Its cyclohexane ring appeared accessible via the classic Diels-Alder reaction.

Retrosynthetic strategy for epibatidine

The synthesis is summarized on the following page. 2-Chloropyridine required as a starting material was available commercially as the acid chloride. Using polymer-supported borohydride, it is fully reduced to the alcohol. Its oxidation to the aldehyde uses polymer-supported perruthenate, and the base for a Henry condensation with nitromethane was provided by an ion-exchange resin. The resulting alcohol was converted to the trifluoroacetate and eliminated with a tertiary amine resin. These steps could also be conducted sequentially in a single vessel using a "reverse tea bag" format, where the reagents were encapsulated in pouches and added to the reaction mixture.

The Diels-Alder reaction of the unsaturated nitro compound required only a 2-fold excess of the diene, and the excess could be removed by evaporation before the initial product was hydrolyzed with TFA to give the cyclohexanone. Reduction of this ketone with polymer-supported borohydride gave a mixture favoring the required *trans* stereochemistry between the nitro group and the alcohol. In conventional synthesis, it would be tempting (or even required) to separate the diastereomers at this stage, but Ley wished to avoid any chromatographic separations in this synthesis, so the problem was addressed later in the sequence. The alcohol was converted to the mesylate and the nitro group was reduced to the primary amine using nickel salts and polymer-supported borohydride, which likely generates a nickel hydride (like Raney nickel) *in situ*.

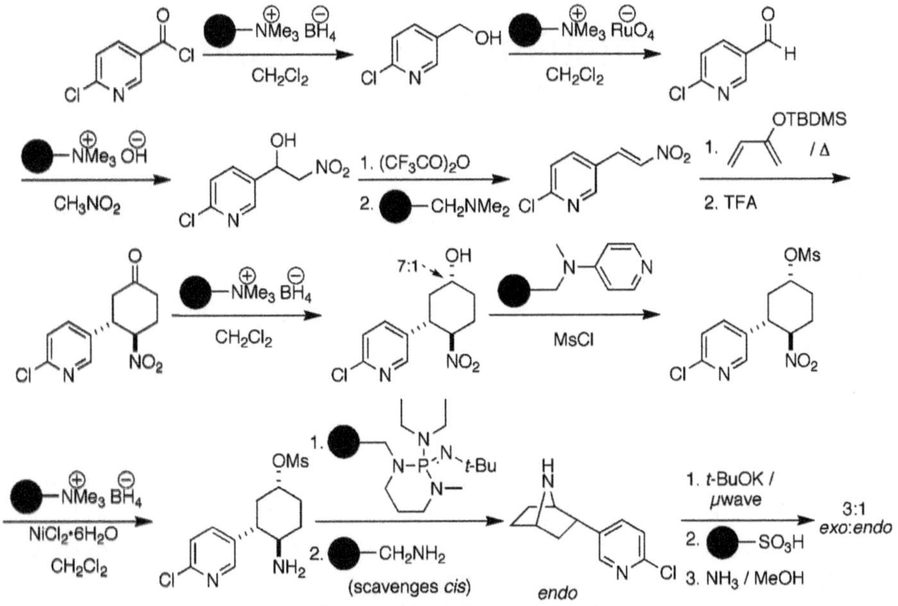

10-Step epibatidine synthesis

The ring closure of the amino-mesylate was accelerated by a polymer-supported phosphazene base. The *cis* diastereomer of the amino-mesylate is unable to cyclize, but could be captured onto (aminomethyl)polystyrene resin to remove it from the reaction mixture. Through this stage, the route produces the *endo* configuration, epimeric to the natural product. This stereochemistry is rectified by a base-promoted equilibration to the thermodynamically more stable *exo* configuration using heating by microwave irradiation. Removal of reagents and by-products of this step was accomplished by absorption of the target compound, which is of course a basic amine, onto an acidic ion-exchange resin, elution of the neutrals, and then elution with ammonia.

Mappicine

Curran has demonstrated the utility of his fluorous synthesis methods in the preparation of derivatives of a polycyclic aromatic natural product, mappicine. First, the overall synthesis will be described. A complex pyridine is converted to an organometallic derivative that adds to an aldehyde. The resulting alcohol is protected with a fluorous silyl group. Two simple steps convert the pyridine to the pyridone, which is *N*-propargylated. This intermediate undergoes a radical cascade cyclization with a phenyl isonitrile in the presence of hexabutylditin as the initiator to form the targets.

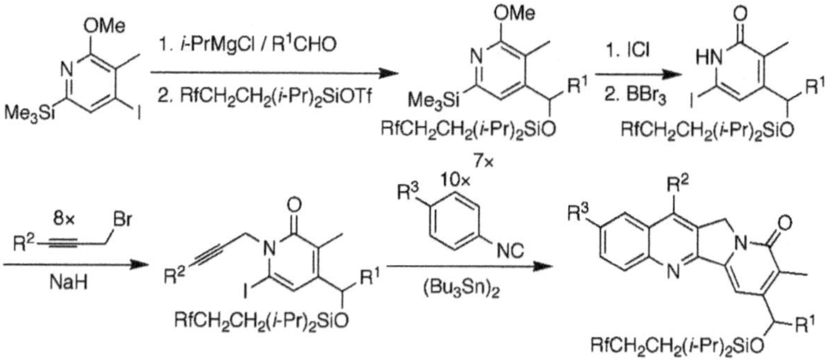

Preparation of mappicine analogues

Diversity is introduced into this synthesis at three points. Varying aldehydes can be used in the first addition reaction, varying alkyne substituents can be used on the propargyl bromide, and varying groups can be substituted onto the aryl isonitrile. The aldehyde diversity is paired with another type of diversity in the protecting group that enables the synthesis to be performed on intermediate mixtures while still permitting the final products to be obtained as pure compounds.

The aldehyde addition products are derivatized with fluorous silyl groups bearing a homologous series of fluoroalkanes of 3-10 carbons. These compounds can be mixed and used as such in the conversion to the pyridone. This "pure mixture" is then split into 8 portions for alkylation with individual propargyl bromides. The chemical and physical properties of the 7 compounds in each of these mixtures should be very similar in the absence of an influence, such as a fluorous phase, that can distinguish them based on fluorine content. Each of these 8 reac-

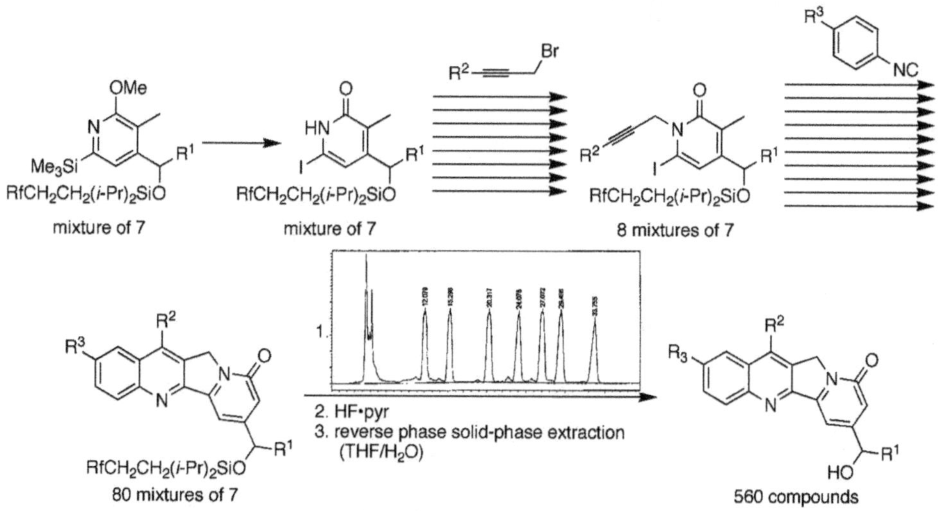

Preparation of a 560-member library

tion products is further split into 10 portions for cascade cyclizations with 10 different isonitriles. At this stage, each of these 80 mixtures is composed of 7 compounds. Each can be rectified to its constituents with fluorous HPLC, and the individual fractions can be desilylated. The product of the deprotection reaction is purified by solid-phase extraction, in which the target compound is not retarded by a reverse-phase sorbent.

Dysidiolide

Complex compounds have also been synthesized on the solid phase. Waldmann developed preparations of dysidiolide derivatives, which are marine natural products that inhibit protein phosphatases. A particularly imaginative feature of this work is the ring-forming release method.

A solid-phase synthesis support **A** was prepared from Merrifield resin by chloride displacement, deprotection of the THP group, and oxidation. Any molecule can be loaded onto the resulting aldehyde resin by Wittig reaction. In this case, ethylidene triphenylphosphorane is alkylated with a complex, scalemic iodide incorporating one of the cyclohexane rings of the target, then deprotonated. Wittig reaction of this ylide with resin **A** gives a mixture of alkene diastereomers **B**.

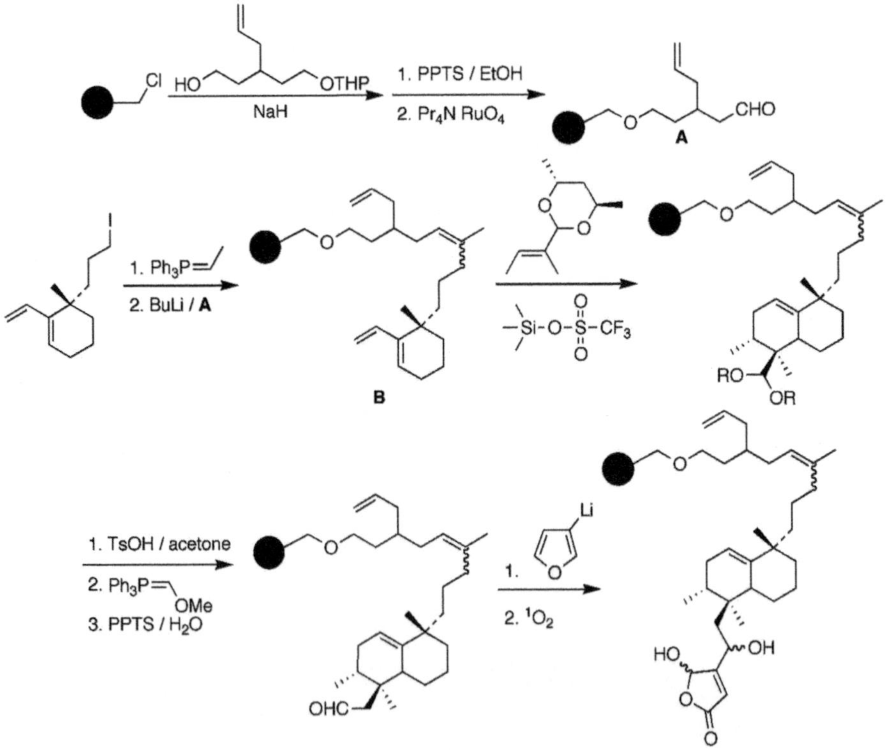

Solid-phase synthesis of dysidiolides

The Diels-Alder reaction of an acetal of the α,β-unsaturated aldehyde tiglaldehyde is promoted by trimethylsilyl triflate at low temperature. This step involves the creation of two new stereogenic centers relative to an existing stereogenic center, so a chiral acetal is used to control the stereochemistry. The major *endo* product shown constituted 87% of the 4-component isomer mixture. Hydrolysis of the acetal and chain extension of the aldehyde set up the addition of a 3-furyl organometallic. This step, expectedly, does not control stereochemistry relative to the three existing stereogenic centers and therefore must create a diastereomeric mixture. The furan ring is then oxidized to the hydroxybutenolide by singlet oxygen.

The skeleton of the target compound was now complete, and ring-forming metathesis released it from the support. Shown is one diastereomer produced, 6-*epi*-dysidiolide. This release strategy bears further discussion. Shown on abbreviated structures is the process by which release occurs. Initial metathesis of the ruthenium vinylidene with the terminal alkene in the linker is much faster than with any of the trisubstituted alkenes in the target molecule. Once the vinylidene is formed on the linker, it can undergo ring-forming metathesis only with the nearby alkene. Once the cyclopentene is formed on the support, the vinylidene derivative of the product can undergo cross-metathesis with another terminal alkene to produce the dysidiolide derivative and start another catalytic cycle.

6-epi-*dysidiolide*

Release of dysidiolides by ring-closing metathesis

This method neatly circumvents a problem that might otherwise befall this synthesis. Metathesis of internal alkenes is much slower than with terminal alkenes. While removal of the product from the support by cross-metathesis with ethylene might be conceivable, it might not be practical. Therefore, to initiate the metathesis process in an intermolecular sense, a reactive terminal alkene is used. Further reactions in the cascade are intramolecular and therefore less subject to the steric limitations of metathesis.

It should be emphasized that this ring-forming release process differs from several related processes earlier discussed. One of the advantages of those release methods was that reactants that did not successfully complete the synthetic sequence could not undergo release, meaning that the cleavage of final products also constituted purification from failure products. In this case, that principal does not apply. Intermediates that failed during the synthesis could still be released in the final step.

This synthesis also involved several steps that were conducted at low or elevated temperatures (-30 °C → 78 °C), further supporting the idea that any reaction that can be performed in solution can be performed on solid phase.

Additional Reading

Brohm, D.; Philippe, N.; Metzger, S.; Bhargava, A.; Muller, O.; Lieb, F.; Waldmann, H. Solid-phase synthesis of dysidiolide-derived protein phosphatase inhibitors. *J. Am. Chem. Soc.* **2002**, *124*, 13171-8.

Habermann, J.; Ley, S. V.; Scott, J. S. Synthesis of the potent analgesic compound epibatidine using an orchestrated multi-step sequence of polymer supported reagents. *J. Chem. Soc., Perkin Trans. 1* **1999**, 1253-55.

Luo, Z.; Zhang, Q.; Oderaotoshi, Y.; Curran, D. P. Fluorous mixture synthesis: a fluorous-tagging strategy for the synthesis and separation of mixtures of organic compounds. *Science* **2001**, *291*, 1766-9.

Zhang, Q.; Rivkin, A.; Curran, D. P. Quasiracemic synthesis: concepts and implementation with a fluorous tagging strategy to make both enantiomers of pyridovericin and mappicine. *J. Am. Chem. Soc.* **2002**, *124*, 5774-81.

Zhang, W.; Luo, Z.; Chen, C. H.-T.; Curran, D. P. Solution-phase preparation of a 560-compound library of individual pure mappicine analogues by fluorous mixture synthesis. *J. Am. Chem. Soc.* **2002**, *124*, 10443-10450.

Problems

1. Write a mechanism for the final step in the mappicine synthesis.
2. Provide intermediates for the final transformation of the dysidiolide synthesis.
3. How would you prepare the following solid-phase reagents?

The Big Picture

Many themes and principles have emerged from the practice of combinatorial chemistry over the past decade. These are some of the "big picture" items that are worth retaining from even a short book such as this one.

- A very powerful concept in finding molecules with a particular function is *selection*, the ability to destroy or otherwise inhibit the survival of undesirable molecules, and to enhance the survival of desirable molecules into a next generation of molecules. Generally, selection requires molecules that can be replicated.

- Less powerful but more often applicable, *screening* involves examining molecules in an assay and requires the experimenter to make a decision as to which molecules are taken forward in the development of compounds with desired properties.

- The central dogma of molecular biology is: DNA makes RNA makes protein. Conversion of DNA to RNA is transcription, and conversion of RNA to protein is translation. This paradigm defines the biological methods to create peptide sequences from DNA sequences.

- Chemistry took on a new focus/challenge in combinatorial chemistry. Instead of making large amounts of one compound, as has been tradition, combinatorial chemistry aims to make small amounts of many compounds. The determination of structure of those compounds must be very simple, or ideally unnecessary. Methods to encode the structure of a compound based on physical or chemical properties were therefore developed. A molecular structure could also be maintained based on the attachment to a particular bead, or to a particular microscopic location on a surface.

- Miniaturization is necessary to generate and maintain large number of compounds, each in small amounts. A solid-phase synthesis bead or microscopic location on a surface is far smaller than any reaction vessel ever devised, but can play roles similar to reaction flasks.

- Synthetic chemists have long been aware that the purification of reaction products is often far more time-consuming than performing a reaction. Methods to accelerate purification significantly enhance the throughput in synthetic chemistry. The purification of compounds is most easily performed by phase separation. At least five phases are available for use in purification: the solid phase, the gas phase, and, in the liquid phase, the aqueous, organic, and fluorous phases. Use of phase separation in product purification significantly enhances automated methods to be used in synthesis.

- Most if not all organic reactions can be adapted to solid-phase methods. Likewise, many of the techniques that are used for analytical characterization of reaction products are adaptable, with some effort, to solid-phase synthesis.

- Solid-phase synthesis is a useful method to separate in space reactions that might otherwise occur in free solution. However, the idea of site isolation on solid phase is not necessarily general. The ability of different functionalized sites on the same polymer bead to interact is related to the loading and is analogous to concentration in solution-phase reactions.

• Split/couple/mix solid-phase synthesis is a powerful method to create a mixed library of compounds in which each synthesis bead bears but one type of molecule.

• There are three primary ways to accomplish the preparation of a library of compounds where each is equally represented (which is often needed to obtain reliable assay data).

1. Split/couple/mix solid-phase synthesis
2. Isokinetic mixtures of reagents (solution or solid phase).
3. Equimolar amounts of unique and pooled reagents (primarily solution phase).

• A powerful method of simultaneously synthesizing a compound on solid phase and preventing failures in the synthesis from contaminating the product is ring-forming cleavage.

• The concept of encoding the structure of a molecule arises repeatedly in combinatorial chemistry. This generally entails a physical link between a molecule that has the desired activity that is being tested for and a molecule that encodes its structure. This paradigm applies to peptides on phage as well as to the Still tag encoding. Molecular coding generally uses binary codes, while biological encoding uses the quaternary (base 4) codes of DNA.

• Multi-component reactions are powerful in bringing together compounds from several different molecular classes, but because they do this in one step, are not useful in the split/couple/mix combinatorial library methods.

• The concept of chemical space that has emerged from library design efforts provides a useful way of thinking about molecular diversity and selecting molecules with diverse properties.

• Arrays of molecules can be prepared on surfaces using masks that prevent some areas of the surface from reacting to create product molecules. Arrays can also be prepared by delivering specific reagents to sites on a surface with miniaturized printing jets. The most useful method to detect a desired molecule or outcome on such an array is optical imaging.

• Miniaturization of combinatorial libraries to the level of a single bacteriophage or even a single RNA molecule enables molecules with binding activity to be directly enriched from large random libraries by immobilization to the target, effectively constituting a phase separation.

Answers to exercises

1. Nature: The Original Combinatorial Chemist

1. Ten variable exons means 2^{10} combinations, or 1024.

2. The new mRNA sequence is AUG GGA ACA AGU AGG A (this last A is irrelevant), which is translated into Met Gly Thr Ser Arg, which compares to the earlier sequence Met Glu Gln Val Gly.

3. DNA - $4^{60} = 1.3292 \times 10^{36}$; peptide - 60 nt = 20 triplets = 20 amino acids $\Rightarrow 20^{20} = 1.0486 \times 10^{26}$.

4. Self-replication enables future generations of a population to come into being. Mitosis is an example. Selective pressure causes some members of a population to fail to survive into future generations. Antibiotic selection of resistant bacteria is an example.

5. For the formula N^I, $N = 4$ and $I = 4$, so there are 256 different oligomers.

2. Synthetic Peptide Libraries

1. SI, TI, CI, SV, TV, CV.

2. It is impossible to know the full sequence without knowing the mass of the M^+ ion. However, the differences starting at the N-terminus identify the sequence HPQ.

3. For the formula N^I, $N = 18$ and $I = 4$, so there are 18^4 or 104,976 different peptides.

4. We will assume that the library remains on the beads, because they are the same if removed from them. Assume l is the length of the sequence synthesized. Considering the sequence(s) on an individual bead, in the split/couple/mix method, each bead will bear only **one** sequence (including stereoisomers). Using the isokinetic mixture method, 40 possibilities can be added at each position of the peptide, and all sequences would be present on the same bead. Assuming the number of functionalization sites on each bead is 2×10^{14}, a very high likelihood that **each** peptide is present at least once on each bead (Poisson mean = 10) would require that peptides be made no longer than 8 residues ($40^8 = 6.5 \times 10^{12}$, which is $< 2 \times 10^{13}$). Each bead would bear a mixture of 2^8 diastereomers.

5. The crude purity must be no greater than the maximum overall yield, calculated as $(0.98)^{10} = 0.82$, or 82%.

6.

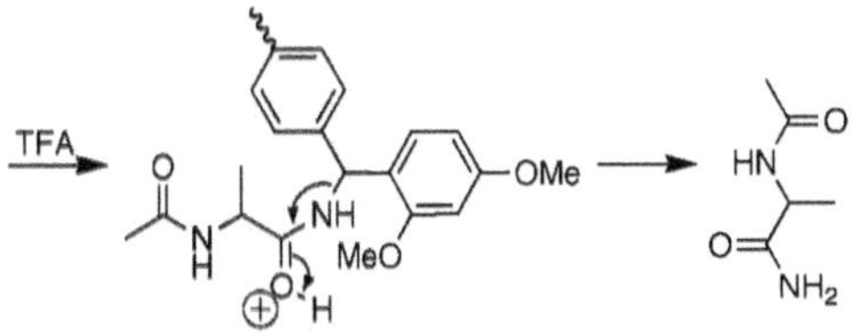

7. For the formula N^l, $N = 3$ and $l = 5$, so there are 3^5 or 243 different peptides. The masses that these amino acids contribute to a peptide are: R - 156; K - 128; N - 114. Therefore, the sequences of the peptides are: 1x - NNRK; 3x - NRRN; 5x - KRRN; 7x - NRRK. Since 1215 beads were used to make the library, the Poisson mean is 5, and valid hits should appear on average five times. The three sequences that are real are therefore NRRN, KRRN and NRRK, and the consensus binding sequence is RR.

8. We will assume that the (R) and (S) forms of a particular amino acid will couple at the same rate. (This is not strictly true, since the relationship between these enantiomers in coupling to a growing peptide chain that is itself chiral is diastereomeric.) Therefore, we will make a mixture that is 50% of the (R) amino acids and 50% of the (S) amino acids. Within each enantiomeric series, the proportion is defined by the inverse of the reaction rates. For example, the valine concentration must be larger than the lysine concentration to make up for its slow rate, so:

$$2V = 3K$$

$$2V = 5F$$

$$3K = 5F$$

A boundary condition is that $V + F + K = 0.5$ (in one enantiomeric series). Doing the algebra, the overall mixture will be:

24.2% (R)-Boc-Val	24.2% (S)-Boc-Val
16.1% (R)-Boc-Lys	16.1% (S)-Boc-Lys
9.7% (R)-Boc-Phe	9.7% (S)-Boc-Phe

9. It is impossible to know the full sequence without knowing the mass of the M^+ ion. However, the differences starting at the N-terminus identify the sequence ANMFX.

3. Supports, Linkers, and Reagents for Peptide Synthesis

1. Their sensitivity to acid - supports for Boc synthesis must be stable to TFA used in the removal of the amine protecting group, while supports for Fmoc synthesis can be cleaved with TFA to make removal of synthesized peptides easier than in Boc chemistry, which uses HF.

2. It is tempting to think that the long PEG chains of TentaGel will make the reaction more likely. However, this intrabead reaction will be concentration controlled. On the bead that means higher loading, and therefore it will occur more readily with the (aminomethyl) polystyrene.

4. Supports and Linkers for Small Molecule Synthesis

1.

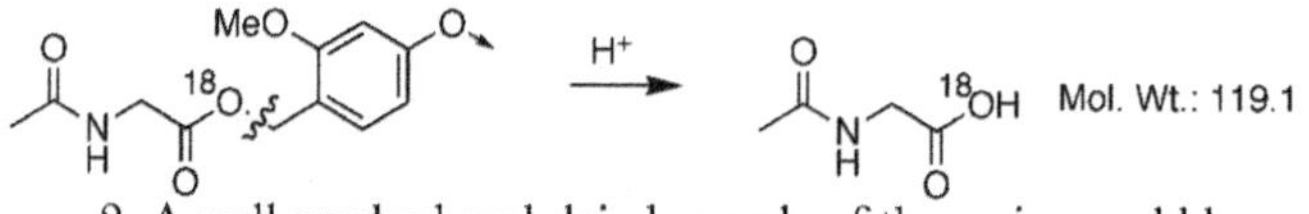

2. A well-washed and dried sample of the resin would be weighed and treated with TFA. The eluent would be evaporated and weighed. To be certain that all of the nonanol had been removed, the process would be repeated. The sum of the released nonanol samples would be used to calculate loading.

3.

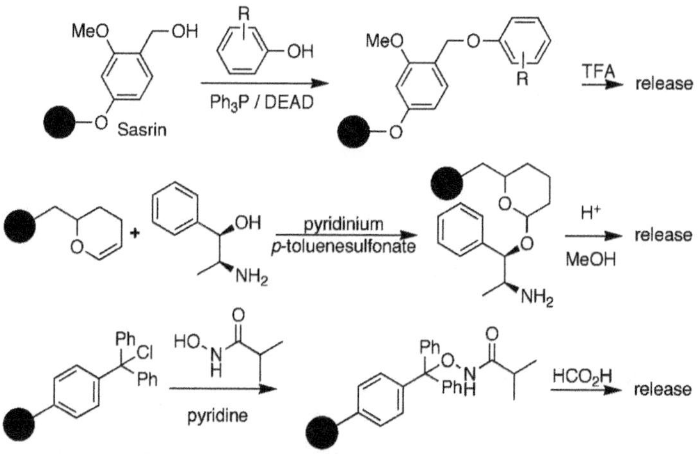

5. Encoded Combinatorial Chemistry

1. For the formula $2^n - 1$, $n = 4$, so there are 15 different code possibilites.

2. Consider a triazine acid that begins as all ^{14}N, ^1H. There are five positions that can incorporate the heavier isotope (^{15}N, ^2H) of each atom, meaning that the isotopically labeled compounds are M, M+1, M+2, M+3, M+4, and M+5; that is, the number is 6. If 3 are used to encode the first amino acid and 3 for the second amino acid, the number of unnatural amino acids is $2^3 - 1$, or 7 (49 dipeptides). If 2 are used for the first amino acid and 4 for the second, it is 3 for the first and 15 for the second (45 dipeptides). It is actually possible to get even more sophisticated by using specific mixtures of the isotopically labeled compounds as the actual codes, so it is not just the presence but the *ratios* of the M, M+1, etc., peaks that determines the code readout.

3. This synthesis will use 3 tyramines and 15 each of the amines and aldehydes, maximizing use of the binary coding molecules and giving a library of 675 molecules. The support will be derivatized with the Holmes photolinker. A total of 4600 beads will be used, meaning that the Poisson mean will be ~7. The scheme on the following page will be used for the synthesis. The beads will be aliquoted 24 at a time into 96-well, filter bottom plates, requiring 2 plates. The plates will be irradiated for a short time to release the 24 compounds/well. The mixture will be eluted with methanol and the solvent evaporated. These wells will be tested and the active wells identified. The beads from four active wells will be distributed into one 96-well plate, it

will be irradiated, single compounds will be eluted with methanol, the solvent will be evaporated, and each of these single compounds will be tested. The single beads in the active wells will be decoded.

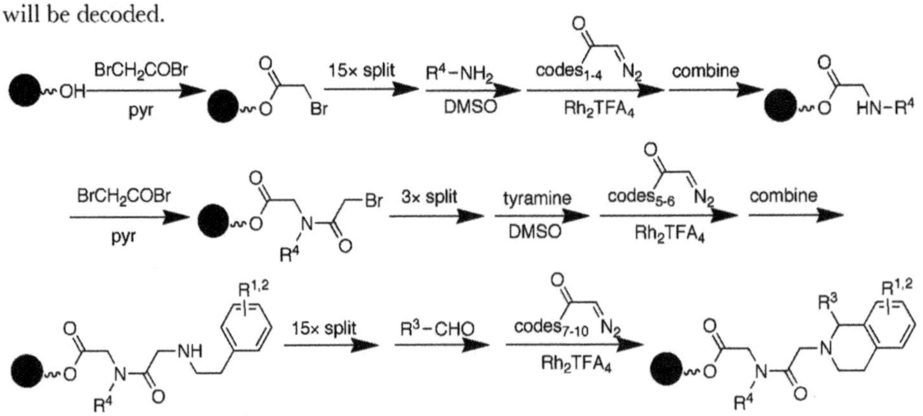

6. Directed Sorting

1. A total of 9216 compounds will be prepared. Therefore, in the first and last diversity-introducing step, the number of Kans in each reagent will be 9216 ÷ 24, or 384. In the second diversity-introducing step, the number of Kans in each reagent will be 576.

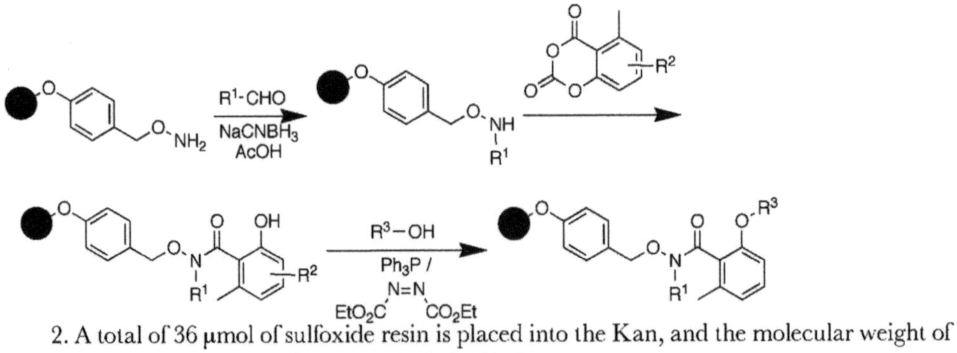

2. A total of 36 μmol of sulfoxide resin is placed into the Kan, and the molecular weight of the product is 250 g/mol, so the theoretical yield is 9 mg.

3. The obvious answer is 2^{144}, but this neglects the fact that the grid must be oriented to be read unambiguously. The grid can be oriented by setting the upper corners to 1 and the lower corners to 0, leaving 140 bits to code the NanoKans. $2^{140} = 1.39 \times 10^{42}$.

7. Unnatural Oligomers for Library Synthesis

1.

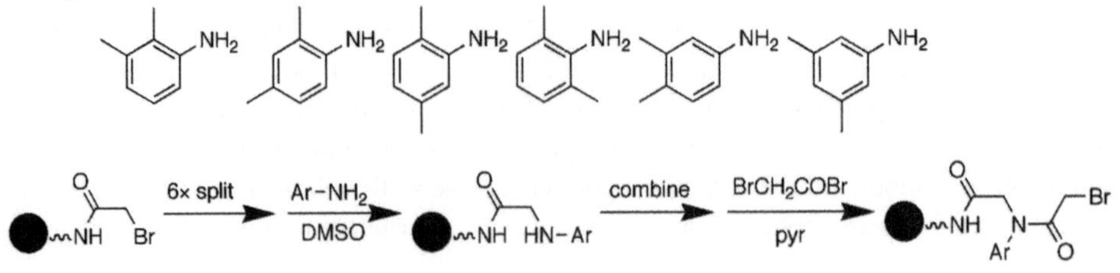

2. The bond that is broken by the protease is replaced by a P-O bond of the phosphonate.

8. Analytical Methods for Solid-phase Synthesis

1. The linker shown below will be cleaved photochemically to give a conjugate of the NSAID to a readily protonated amine that will sensitize the compounds for electrospray mass spectrometry analysis. Also, with two bromine atoms on the linker, compounds that include it will give a distinctive 1:2:1 molecular ion pattern. Removal of the NSAID from the support can be performed with conventional acid treatment.

2. a. Single bead FTIR microscopy would be a good method because of the distinctive IR absorption of a terminal acetylenic C-H bond at about 3300 cm^{-1}. b. Gel-phase carbon NMR should prove useful for this (aminomethyl)polystyrene resin provided that ^{13}C-bromoacetic acid is used. c. With the PEG chain, HR MAS proton NMR should be a good technique for this support. The starting material has a distinctive aldehyde hydrogen and the product has a distinctive *tert*-butyl group.

3. Gel-phase carbon NMR could be used to monitor the change of chemical shift of the α carbon if the labeled bromoacetic acid were substituted for bromoacetyl bromide.

4. Gel-phase NMR would be a useful technique to monitor the nucleophilic aromatic substitution of fluorine. The ketone reduction could be monitored by HR MAS proton NMR because the multiplicity of the methyl group will change, and this could be discerned either directly or by two-dimensional *J*-spectroscopy.

9. Supported Solution-phase Synthesis

1. Since all of the failure sequences will be highly fluorous (C_9F_{19}), passing the crude reaction mixture through a plug of fluorous silica gel should retain all of the capped products, but since the uncapped peptide has little affinity for the stationary phase, it should be readily eluted. It is unlikely with only 19 fluorines that the failure sequences would partition exclusively into a fluorous liquid phase.

2. Because the starting material is fluorous, it can dissolve in the FC-72 and be transported over to the MeOH phase. It partitions partly into the MeOH because of its organic part. There, the fluoride deprotects it to give the cinnamyl alcohol plus the fluorous silyl fluoride. Because it is mostly fluorous, it preferentially partitions back into the FC-72, leaving only the alcohol in the MeOH.

10. Solution-phase Parallel Synthesis

1. Amines are treated with excess phenylisothiocyanate. An excess of the fluorous amine is then added, converting all of the phenylisothiocyanate to the fluorous urea. Both fluorous

amine and urea are removed by partitioning into FC-72. The organic solution of the phenyl-thiourea is evaporated.

2.

3.

4. Partitioning between benzene and FC-72 will give the aniline in the organic phase and the fluorinated tertiary alcohol in the fluorous phase.

11. Multi-component Reactions

1.

2.

3.

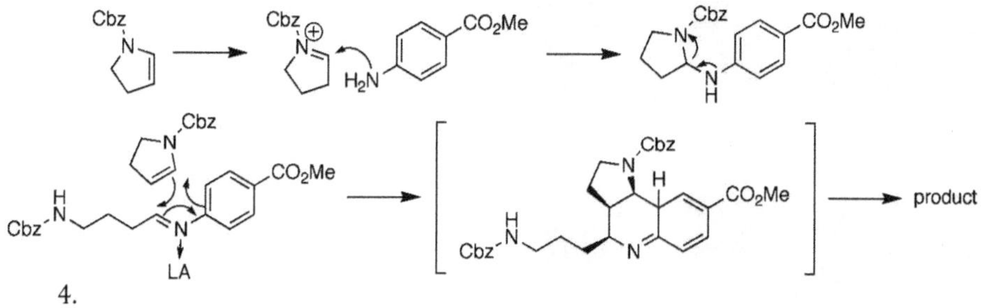

4.

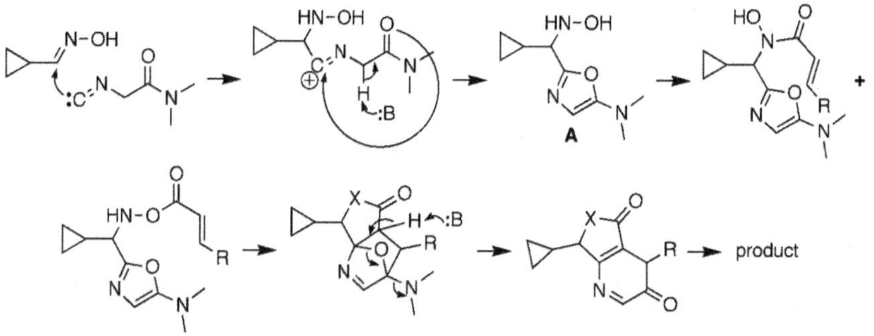

12. Chemical Informatics, Diversity, and Library Design

1. The isocyanoamide must be prepared from the commercially available isocyanoester by hydrolysis and coupling with dimethylamine. The oxime must be prepared by treatment of commercially available cyclopropane carboxaldehyde with hydroxylamine. There are 22 different *trans*-unsaturated acid chlorides that could be used in the synthesis. The library size would be 22 × 2, or 44. 3-(2-thienyl)acryloyl chloride comes from Ubichem.

2. The compound that does not obey Lipinski's Rules is novobiocin ($C_{31}H_{36}N_2O_{11}$).

13. Nucleic Acid Microarrays

1. For the formula N^l, $N = 4$ and $l = 12$, so there are 16.7 million different sequences prepared in 48 steps.

2. While in principle there are 4^{10} sequences, note that these are not directional oligomers (each sequence will be symmetrical). Therefore, there are $4^{10} ÷ 2$, or 524,288 sequences.

14. Combinatorial Materials Chemistry

1. With four building blocks, they must be used in two rounds of synthesis with quaternary masks to make binary mixtures. In general, we expect 4^n compositions from $4n$ steps. However, in this case because we must use the same building blocks twice, there will be duplicates of several compositions. That is, In/Sn is the same as Sn/In, and Sn/Sn is produced in two squares. We would normally expect 64 compositions, but in this case there are only 10 possible combinations.

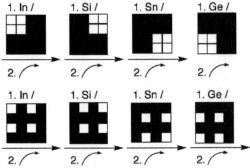

15. Combinatorial Catalyst Discovery

1. If a catalyst can promote a reaction in the absence of a chiral influence, it will produce a racemic product. In order to produce a product with the highest possible enantiomeric excess, this reaction must be much slower than the reaction of a catalyst that does have a chiral influence. The chiral influence is provided by the chiral ligand, which must also make the reaction faster.

2. They are exothermic, meaning that any detector for the heat of reaction should reveal catalysis.

3. A series of metal ions and a series of Y ligands would be chosen that might be expected to combine to promote the reaction. Because of the longer conjugation length of the product, it should exhibit significant UV/visible absorption red-shifted from the starting materials, so simple optical spectroscopy should provide a reliable continuous assay. To determine the "background" rate, the reactants would be mixed alone and with only individual ligands and with only individual metal ions. The fastest of these reactions provides the control for the ligand-accelerated reaction. Each metal ion would then be mixed one at a time with all ligands, and each ligand would be mixed one at a time with all metal ions. These X + Y mixtures would be assayed for acceleration of the reaction above the fastest background reaction. The fastest X mixtures would identify the best metal ion, and the fastest Y mixtures would identify the best ligand.

16. Peptides on Phage

1. It is tempting to guess that because the number of independent recombinants in this library is far smaller than the computed diversity [$20^{10} \sim 10^{13} >>> 10^9$], there was not even one binding peptide in the phage library that was actually screened, though we might have expected there would be. However, given that we took a fraction of the phage and tried to infect, they should still give plaques. Therefore, the more likely possibility is that peptide sequences that can bind to maltose are specifically unable to infect *E. coli*. They must create a pIII that is unable to bind to the F-pilus. One way to get around this would be to engineer the peptide into the pIX or pVI proteins, which are not involved in infection but are still present in low copy number.

2. The sequence is: VTSPDSTTGAMA. The large number of hydroxyl-containing amino acids suggests the primary binding mode is Lewis acid-Lewis base.

3. Split/couple/mix is needed only when one wants to make one bead/one compound libraries for on-bead screening. When the products are to be removed from the support, split/

couple/mix and mixture synthesis give the same library mixture, but isokinetic mixture synthesis is much more efficient (a single coupling per position rather than four).

4.

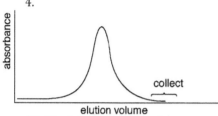

elution volume

17. Nucleic Acid Selection

1. With two pyrimidines, the number of sequences is 2^{40} or 1.1×10^{12}. For 10^{15} sequences, the length of a random region of two pyrimidines can be determined by the equation $2^n = 10^{15}$. The solution for n is 49.8. That should be rounded down to 49, which would give a Poisson mean of 1.78. A library with this Poisson mean would have a significant fraction of missing sequences. A more realistic length of the random region would be 47, which increases the Poisson mean to 7.1.

2. This material could be prepared by solid-phase oligonucleotide synthesis, but it must necessarily be in the unnatural enantiomeric series to resist degradation by endogenous enzymes. The unnatural enantiomers of A, G, T, and C would therefore have to be prepared, protected, and converted to their phosphoramidites.

3. Metal-binding affinity matrices would be prepared and complexed with mercury or silver ion. The RNA population would be passed down the mercury column and the *last* eluting 0.001% of the library isolated. This sample would be subjected to PCR and conversion to single-strand RNA. This population would be passed down the silver affinity column and the *first* eluting 0.001% of the library isolated. This sample would be subjected to PCR and conversion to single-strand RNA. These steps would be repeated alternately until the majority of the RNA library *does* bind to the mercury column but *does not* bind to the silver column.

4.

18. Complex Combinatorial and Solid-phase Synthesis

1.

2.

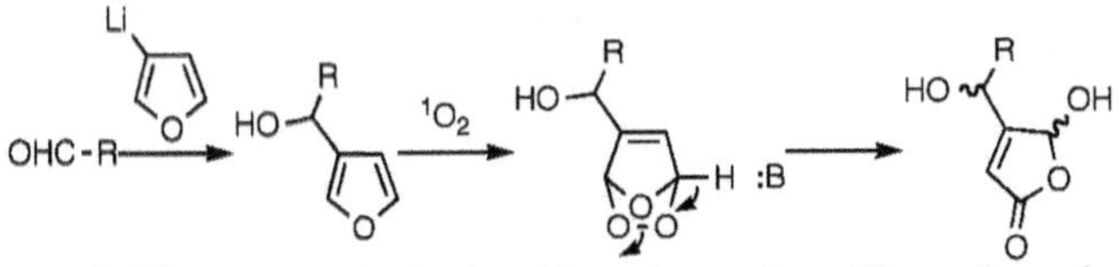

3. We convert the hydroxide resin to the sulfate salt so that when barium permanganate is added, the very stable barium sulfate is formed, driving generation of the permanganate salt of the resin. If the hydroxide resin is treated with hydrogen peroxide, it will be converted to the hydroperoxide salt because hydrogen peroxide is ~10^6 more acidic than water.

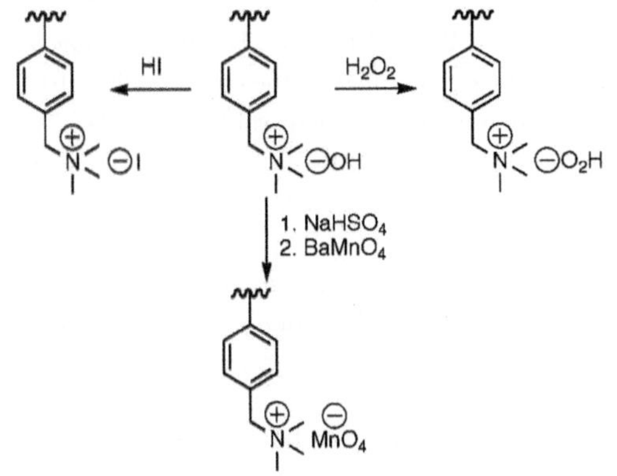

Bibliography

Major research texts on combinatorial chemistry

"Combinatorial Strategies in Biology and Chemistry," A. Beck-Sickinger, P. Weber, Wiley (2002).

"Combinatorial library: Methods and Protocols (Methods in Molecular Biology, v. 201)," L. B. English (Ed.). Humana Press (2002).

"Optimization of Solid-Phase Combinatorial Synthesis," B.Yan, A. W. Czarnik (Eds.), Marcel Dekker (2002).

"Combinatorial Library Design and Evaluation : Principles, Software Tools, and Applications in Drug Discovery," A. K. Ghose, V. N. Viswanadhan (Eds.), Marcel Dekker (2001).

"Solid-phase Organic Synthesis," A. W. Czarnik, (Ed.), John Wiley (2001).

"Solid Support Oligosaccharide Synthesis and Combinatorial Carbohydrate Libraries," P. H. Seeberger (Ed.), John Wiley & Sons (2001).

"High-throughput Synthesis: Principles and Practices," I. Sucholeiki (Ed.), Marcel Dekker (2001).

"Organic Synthesis on Solid Phase: Supports, Linkers, Reactions" F. Dörwald, John Wiley & Sons (2000).

"Combinatorial Catalysis and High Throughput Catalyst Design And Testing," E. G. Derouane (Ed.), Kluwer (2000).

"Solid-Phase Synthesis and Combinatorial Technologies," P. Seneci, John Wiley & Sons (2000).

"Combinatorial Chemistry: A Practical Approach (The Practical Approach Series, 233)," H. Fenniri (Ed.), Oxford University Press (2000).

"Combinatorial Chemistry: A Practical Approach (Methods and Principles in Medicinal Chemistry)," W. Bannwarth, E. Felder (Eds.), Verlag Chemie (2000).

"Analytical Techniques in Combinatorial Chemistry," M. E. Swartz, (Ed.), Marcel Dekker (2000).

"Solid-phase Organic Synthesis," K. Burgess (Ed.), Wiley-Interscience (2000).

"Combinatorial Chemistry In Biology (Current topics in Microbiology and Immunology, 243)," M. Famulok, E.-L. Winnacker, C.-H. Wong (Eds.), Springer (1999).

"Combinatorial Chemistry: Synthesis, Analysis, Screening," G. Jung (Ed.), Verlag Chemie (1999).

"Combinatorial Chemistry and Technology: Principles, Methods, And Applications," S. Miertus, G. Fassina (Eds.), Marcel Dekker (1999).

"Combinatorial Peptide Library Protocols (Methods in Molecular Biology, v. 87)," S. Cabilly, Humana Press (1998).

"Combinatorial Chemistry and Molecular Diversity in Drug Discovery" E. M. Gordon, J. F. Kerwin (Eds.), Wiley-Liss (1998).

"Combinatorial Chemistry," N. K. Terrett, Oxford University Press (1998).

"The Combinatorial Index," B.A. Bunin, Academic Press (1998).

"A Practical Guide to Combinatorial Chemistry," A. W. Czarnik, S. H. Dewitt, (Eds.), American Chemical Society (1998).

"Solid-supported Combinatorial and Parallel Synthesis Of Small-Molecular-Weight Compound Libraries," D. Obrecht, J. M. Villalgordo, Pergamon / Elsevier Science (1998).

"Combinatorial Chemistry: Synthesis and Application," S. R. Wilson, A. W. Czarnik, Wiley-Interscience (1997).

"Combinatorial Peptide and Nonpeptide Libraries: A Handbook," G. Jung (Ed.), John Wiley & Sons (1997).

"Combinatorial Chemistry (Methods in Enzymology, v. 267)," J. N. Abelson, (Ed.), Academic Press (1996).

URLography

http://www.wkap.nl/prod/j/1381-1991 -- Molecular Diversity, an on-line journal

http://www.combichem.net/

http://www.5z.com -- a dynamic database of references in molecular diversity, curated by M. Lebl

and

http://www.5z.com/divinfo/classics.html -- a compilation of classic papers in combinatorial chemistry.

"Analytical methods in combinatorial chemistry." Yan, B. Technomic (2000) - http://www.netLibrary.com/ebook_info.asp?product_id=32562.

"High throughput screening: the discovery of bioactive substances," J. P. Devlin (Ed.), M. Dekker (1997). http://www.netLibrary.com/ebook_info.asp?product_id=12699.

http://www.combinatorial.com/ -- a continuously updated, ongoing Combinatorial Index website.

http://www.CombiChemLab.com -- a review of reaction blocks, devices and workstations for synthesis, purification and analysis, software for library design, management and analysis, and sources for building blocks

http://chemacx.cambridgesoft.com/chemacx/index.asp -- ChemACX

Reviews

General

Hall, D. G.; Manku, S.; Wang, F. Solution- and solid-phase strategies for the design, synthesis, and screening of libraries based on natural product templates: a comprehensive survey. *J. Comb. Chem.* **2001**, *3*, 125-50.

Dolle, R. E. Comprehensive survey of combinatorial library synthesis: 2002. *J. Comb. Chem.* **2003**, *5*, 693-753.

Dolle, R. E. Comprehensive survey of combinatorial library synthesis: 2001. *J. Comb. Chem.* **2002**, *4*, 369-418.

Dolle, R. E. Comprehensive survey of combinatorial library synthesis: 2000. *J. Comb. Chem.* **2001**, *3*, 477-517.

Dolle, R. E. Comprehensive survey of combinatorial library synthesis: 1999. *J. Comb. Chem.* **2000**, *2*, 383-433.

Dolle, R. E, Nelson, K. H., Jr. Comprehensive survey of combinatorial library synthesis: 1998. *J. Comb. Chem.* **1999**, *1*, 235-82.

Weber, L. High-diversity combinatorial libraries. *Curr. Opin. Chem. Biol.* **2000**, *4*, 295-302.

Weber, L.; Illgen, K.; Almstetter, M. Discovery of new multi component reactions with combinatorial methods. *Synlett* **1999**, *3*, 366-374.

Bunin, B. A.; Dener, J. M.; Livingston, D. A. Application of combinatorial and parallel synthesis to medicinal chemistry. *Annu. Rep. Med. Chem.* **1999**, *34*, 267-286.

Dolle, R. E. Comprehensive survey of chemical libraries yielding enzyme inhibitors, receptor agonists and antagonists, and other biologically active agents: 1992 through 1997. *Mol. Diversity* **1998**, *3*, 199-233.

Dolle, R. E. Comprehensive survey of combinatorial libraries of undisclosed biological activity: 1992 through 1997. *Mol. Diversity* **1998**, *3*, 233-256.

Thompson, L. A.; Ellman, J. A. Synthesis and applications of small molecule libraries. *Chem. Rev.* **1996**, *96*, 555-600.

Microarrays

Pirrung, M. C. How to make a DNA chip. *Angew. Chem., Int. Ed.* **2002**, *41*, 1276.

Pirrung, M. C. Spatially-addressable combinatorial libraries. *Chem. Rev.* **1997**, *97*, 473.

Supports

Blaney, P.; Grigg, R. Sridharan, V. Traceless solid-phase organic synthesis. *Chem. Rev.* **2002**, *102*, 2607-2624.

Guillier, F.; Orain, D.; Bradley, M. Linkers and cleavage strategies in solid-phase organic synthesis and combinatorial chemistry. *Chem. Rev.* **2000**, *100*, 2091-2157.

Haag, R. Dendrimers and hyperbranched polymers as high-loading supports for organic synthesis. *Chem.--Eur. J.* **2001**, *7*, 327-335.

Coe, D. M.; Storer, R. Solution-phase combinatorial chemistry. *Mol. Diversity* **1999**, *4*, 31-38.

Nilsson, U. J. Solid-phase extraction for combinatorial libraries. *J. Chromatogr., A* **2000**, *885*, 305-319.

Thompson, L. A. Recent applications of polymer-supported reagents and scavengers in combinatorial, parallel, or multistep synthesis. *Curr. Opin. Chem. Biol.* **2000**, *4*, 324-337.

Library design

Xue, L.; Bajorath, J. Molecular descriptors in chemoinformatics, computational combinatorial chemistry, and virtual screening. *Comb. Chem. High Throughput Screening* **2000**, *3*, 363-72.

Spellmeyer, D. C.; Grootenhuis, P. D. J. Recent developments in molecular diversity. Computational approaches to combinatorial chemistry. *Annu. Rep. Med. Chem.* **1999**, *34*, 287-296.

Agrafiotis, D. K.; Myslik, J. C.; Salemme, F. R. Advances in diversity profiling and combinatorial series design. *Mol. Diversity* **1999**, *4*, 1-22.

Warr, W. A. Combinatorial chemistry and molecular diversity. An overview. *J. Chem. Inf. Comput. Sci.* **1997**, *37*, 134-140.

Analytical techniques

Nemeth, G. A.; Kassel, D. B. Existing and emerging strategies for the analytical characterization and profiling of compound libraries. *Annu. Rep. Med. Chem.* **2001**, *36*, 277-292.

Kassel, D. B. Combinatorial chemistry and mass spectrometry in the 21st century drug discovery laboratory. *Chem. Rev.* **2001**, *101*, 255-267.

Fitch, W. L. Analytical methods for quality control of combinatorial libraries. *Mol. Diversity* **1999**, *4*, 39-45.

Shapiro, M. J.; Wareing, J. R. NMR methods in combinatorial chemistry. *Curr. Opin. Chem. Biol.* **1998**, *2*, 372-375.

Congreve, M. S.; Ley, S. V.; Scicinski, J. J. Analytical construct resins for analysis of solid-phase chemistry. *Chem. Eur. J.* **2002**, *8*, 1768-76.

Catalysis/materials

Engstrom, J. R.; Weinberg, W. H. Combinatorial materials science: paradigm shift in materials discovery and optimization. *AIChE J.* **2000**, *46*, 2-5.

Sun, T. X. Combinatorial search for advanced luminescence materials. *Biotechnol. Bioeng.* **1999**, *61*, 193-201.

Jandeleit, B.; Schaefer, D. J.; Powers, T. S.; Turner, H. W.; Weinberg, W. H. Combinatorial materials science and catalysis. *Angew. Chem., Int. Ed.* **1999**, *38*, 2494-2532.

Reetz, M. T. Combinatorial and evolution-based methods in the creation of enantioselective catalysts. *Angew. Chem. Int. Ed.* **2001**, *40*, 284-310.

Senkan, S. Combinatorial heterogeneous catalysis-a new path in an old field. *Angew. Chem., Int. Ed.* **2001**, *40*, 312-329.

Dahmen, S.; Brase, S. Combinatorial methods for the discovery and optimisation of homogeneous catalysts. *Synthesis* **2001**, 1431-1449.

Pescarmona, P. P.; Van der Waal, J. C.; Maxwell, I. E.; Maschmeyer, T. Combinatorial chemistry, high-speed screening and catalysis. *Catal. Lett.* **1999**, *63*, 1-11.

Index

Symbols

^{19}F NMR 72
^{31}P NMR 72
β-cyanoethyl 108

A

active ester 40, 42, 43
adenosine 144
affinity column chromatography 135
Agilent 113
ampicillin resistance 6
amplification 7, 12, 135, 136, 141, 142, 143
analytical construct 74
antibody 10, 11
aptamer 144, 145, 146, 147
asymmetric catalyst 121, 124, 125, 126
Available Chemicals Directory 102
azatide 67

B

bacteriophage 131, 156
base pairs 3, 8, 111
benzotrifluoride 81
Biginelli condensation 98
borohydride resin 89
building blocks 1-4

C

caffeine 145
capping 17
carbodiimide 42
carbonic anhydrase 54, 55, 56
catechol ether 53
Celite 89, 90, 91
central dogma of molecular biology 7, 8, 131, 155
chemical diversity 149
chemical space 101, 102, 104, 156
chemiluminescent nitrogen detector 76
chlorotrityl 45, 46, 50
clones 7
cloning 143
coat proteins 131
codon 8, 13, 132, 133
compositional spread 119
core 5, 76, 82, 102

D

deconvolution 19, 20, 24, 25, 30, 34, 35, 66, 68, 124, 125
dendrimers 80
depletion assay 68
dihydropyran 47
dimethoxytrityl 49
directed evolution 146
disulfide 10, 134, 136, 138
DNA polymerase 131, 133
double coupling 17
dysidiolide 152

E

Edman degradation 27, 28
electron-capture GC 51, 52
electroporation 134
electrospray ionization 74
ELISA 27
Ellman 46, 50
enantiomeric excess 124, 126, 127, 129
encoding 55, 131, 156
enzyme inhibition 20
epibatidine 149
epitope scan 18, 19, 24, 28
erythropoietin 136
evaporation 149
evaporative light scattering detector 76
evolution 4, 5, 7, 12
excimer 38
expressed sequence tags 4

F

F-factor 131
F-pilus 131, 132
FC-72 80, 81
Fetizon's reagent 90
flower plot 105
Fluofix 82
fluorescent dye 123
fluorous 80, 81, 82, 83, 90, 91, 150
fluorous silyl group 98, 150, 151
four-component condensation 81, 93
front-loaded SELEX 146

G

G-quartet 144
gas chromatography 127
gel phase 71
genes 5, 10, 11
genetic code 7, 8, 9, 13, 132, 133
gene expression analysis 113

Geysen 17, 18, 20, 31, 131
gradient 118, 119
Grignard 81, 88

H

haloaryl ether 53
handles 39
Hantzsch synthesis 98
high-resolution magic angle spinning 72
Houghten 24, 32, 34, 59
hybridization 101, 111, 112, 113
hydroxybenzotriazole 16, 42
hydroxysuccinimide 40, 42

I

infrared 73
iodoacetonitrile 46
ion-exchange resin 86, 89, 149, 150
IRORI 59, 60, 61
IR microscopy 74
isobaric 51
isokinetic mixture 21, 23, 31, 35, 141, 156
isopycnic 25, 122

J

J-resolved NMR 73
jets 113, 119, 123, 156

K

Kaiser oxime 39, 40

L

ligand-accelerated catalysis 124, 125, 130
light-directed synthesis 107, 108, 110, 113
light fluorous 82
lipases 128
Lipinski's Rules 105, 106
liquid chromatography/mass spectrometry 75
loading 37, 38, 39, 41, 47, 49
log P 101, 105

M

malachite green 144, 145
Mannich reaction 98
mappicine 150
mask 107, 109, 110, 115, 116, 118, 119, 120, 156
mass spectrometry 74, 76
Merrifield resin 16, 17, 35, 38, 39, 41, 72, 152
microarrays 107, 111, 113
MicroKan 61, 63
microtiter plate 17, 18, 54

mimotope 19, 20, 24, 25, 31
Mitsunobu reaction 68, 91
mRNA 7, 8, 9, 13
multi-component reactions 93, 97, 99, 156
mutation 12, 113, 146

N

N-substituted glycine 65, 69
NAD 144
NanoKan 61, 62
natural selection 5, 7
ninhydrin 17
NOE 72
nuclear Overhauser effect 71, 72
nucleophilic catalyst 121, 128
nucleotide triphosphates (NTPs) 142, 146

O

oligourea 69
one bead/one compound 23, 27, 31, 71, 79
one gene, one protein 7, 10

P

panning 134, 135, 136
parallel synthesis 85
Passerini reaction 96, 97
PEG (polyethyleneglyco) 41, 79
peptide 3, 4, 94, 122
peptoid 65, 66, 69, 122
Pharmacopeia 53, 56
phosphonate 67, 68
phosphonium 43
phosphoramidite 107, 108
phosphors 117
photochemically removable 108, 109, 110, 112
photolabile 53
photolinker 40
photolithography 107
pins 17, 18, 19, 20, 27
piperidine 15, 17, 18
plasmid 5, 6, 7
Poisson 23, 30
polymer-supported borohydride 149, 150
polymer-supported perruthenate 89, 149
polymer-supported phosphazene 150
polymerase chain reaction 51, 141
polystyrene 37, 41, 45, 47, 54, 127
polyvinylpyridinium perbromide 89
positional scanning 30, 31, 32
principle components analysis 104
promoter 9, 141
protecting group 37
proteins 1, 4, 5, 8, 12

protiodesilylation 47, 48
proton-enhanced fluorophore 129
purines 1
pyrimidines 1

Q

quasi-enantiomers 129

R

racemization 38, 42
Raman spectroscopy 74
reading frame 8, 13
recombination 11
replicative form 131, 139
resin capture 94, 95, 96
reverse transcriptase 142, 146
rhodium trifluoroacetate 54
riboflavin 144, 145
ring-closing metathesis 48, 49, 153
ring-forming cleavage 48, 156
Rink amide 39, 61, 94
Rink ester 39
RNA polymerase 9, 141

S

S-adenosyl methionine 144
safety catch 46
salen 125
Sasrin 45, 50
scaffold 102
scavenging resins 85
screening 155
secondary structure 141, 143, 147
selection 155
SELEX 141, 142, 144, 145, 146, 147, 148
self-encoding 51
silyl resin 47
site isolation 37, 155
Smith 131
sodium cyanoborohydride 90
solid-phase extraction 82, 83, 91, 152
solid-phase reagents 85, 89
solid-phase synthesis 15
solid-phase synthesis resin 60, 61
solution-phase syntheses 149
spatially defined 17
spatial encoding 111
spectrometry 129
split/couple/mix 21, 23, 31, 51, 66, 93, 122, 134, 156
split/couple/pool 21
SPPS 15, 16, 20, 21, 24, 59, 67
Src homology 3 138
Still 52, 53, 56, 57, 58, 122, 156

stochastic 23
Strecker reaction 126
sub-monomer 65
superconductor 116
supported reagents 88
swelling 37, 41

T

tea bag 24, 59, 149
template 99, 102, 103, 104, 133, 142, 146
TentaGel 33, 41, 44, 68, 71, 72
TFA 39, 45, 46, 48, 61
theophylline 145
thermography 121, 122
traceless linkers 47
transcription 7, 8, 9, 131, 141, 155
transfection 133
transformation 5, 6, 7, 134, 139
translation 7, 8, 131, 155
transponder 59, 60, 61
triaminotriazine 102
trifluoroacetic acid (TFA) 15, 17, 29
trifluoroethanol 81
trimethylamine-N-oxide 89
trityl 45, 46
truncation sequence 17, 29
tyrphostins 61

U

Ugi 81, 82, 93, 94, 95, 96
ultrafiltration 80
uronium 43

V

vascular endothelial growth factor 146
vector 5, 6
virtual library 103, 104, 105

W

Wang resin 39

X

X-ray 117

Y

YGGFL 26, 28, 67, 69

Printed and bound by CPI Group (UK) Ltd, Croydon, CR0 4YY

03/10/2024

01040399-0020